From Where Comes the Universe?

A Guide for the Lay Person

to the

Theoretical Physics of Empty Space

By

Dennis Morris

(September 2015)

© Dennis Morris

All Rights Reserved

Published by: Abane & Right

31/32 Long Row

Port Mulgrave

Saltburn

TS13 5LF

01947 840707

September 2015

Contents

Is This Book for You?...1
An Overview 4
    Empty space:..5
    An overview of empty-space: 7
We Begin with 1-dimensional Numbers.................11
    We begin: 11
    1-dimensional space:......................................13
    A taste of what is to come:...........................16
    What are numbers?..17
2-dimensional numbers:..20
    Determinants and distance functions: 21
    A little history: 22
    Rotation:...24
    A little trigonometry: 25
    The polar form and rotation matrices:............28
    Summary of this chapter: 30
Fibre Bundles 32
Strange Rotations 35
    Spinor rotations:..35
    Space-time rotations:......................................36
    Distance in 2-dimensional space-time: 39
    What about higher dimensional rotations?.....40
    Stranger and stranger: 48

Contents

- Summary of this chapter: 49
- A Catalogue of Spinor Rotations 52
  - How many finite groups are there? 56
  - The physical universe: 57
  - A note: 60
  - Another taster: 61
- The Special Theory of Relativity 62
  - Summary: 70
  - An addendum – the expanding universe: 70
- Super-imposition 72
  - Super-imposition of spinor spaces: 73
  - Super-imposition - 2-dimensional: 74
  - Super-imposition - 4-dimensional: 75
  - The expectation distance function: 78
  - A flat tangent space: 78
  - But there is no affine connection: 81
  - What about the quaternions? 82
  - What about the 8-dimensional spinors? 83
  - What about the 3-dimensional spinors? 85
  - Conclusion: 86
  - Summary: 86
  - An important point: 87
- Fibre Bundles and QFT 88
  - Fibre bundles again: 89

- The photon field: 90
- The shot-put force: 91
- The quaternions and the weak nuclear force:. 92
- The strong nuclear force: 93
- The A_3 spinors spaces: 95
- What about the 3-dimensional spinors? 96
- Summary: 97

General Relativity 98
- The first postulate of GR: 98
- The second postulate of GR: 99
- The third postulate of GR: 100
- What GR does not explain: 101
- GR gravity compared to the other forces: 102
- The structure of GR space-time: 103
- Summary: 105

The Affine Connection 106

Some Bits 109
- Classical electro-magnetism and GEM: 109
- Meanwhile, back in quantum physics: 110
- And super-symmetry: 110
- Wot! no gravitons: 115
- Breakdown of parity: 116

Concluding Remarks 117
- Two types of physics: 118

A farewell: 119
A philosopher's cogitation: 119
Other Books by the Same Author 121
Index 127

Introduction

Is This Book for You?

This book is a non-mathematical description of our present understanding of the empty space and time of our universe. After centuries with only little progress in understanding empty space and time, there have been many sudden and recent advances in this area. Your author is at the forefront of these advances. As with all areas of physics, there is much mathematics in this area, but your author opines that the mathematics need not be a barrier to the interested lay person. This book presents those recent advances in terms which the lay person can understand.

By non-mathematical, your author means without 'hard mathematics'. We cannot describe our universe without some use of mathematics; we need numbers and concepts like angles and lengths just as we need to be able to read and write. Since anyone who can read is familiar with numbers, angles and lengths, your author assumes such familiarity. The reader is not expected to be on friendly terms with tensor calculus or Lagrangian field theory, and we do not need her to be so. The mathematical level of this book is no more than the understanding obtained by an average school pupil by the age of sixteen.

Although a detailed knowledge of 'hard mathematics' is essential to understand the mathematical proofs which put the stamp of correctness upon our understanding of the universe, if the reader will simply take your author's word that these things have been proven, then we can manage very well without the detailed 'hard mathematics'. Although one could neither become a theoretical physicist nor calculate the outcome of experiments without the requisite detailed mathematical knowledge, at a conceptual level it really is quite easy to gain a deep understanding of this area of modern theoretical physics without the 'hard mathematics'. This does not mean that the understanding will be 'dumbed down'. There are concepts produced by the mathematics which are very challenging.

Many of the concepts of modern theoretical physics are very challenging; this is what makes it so interesting. The reader ought not to be dismayed because she finds the concepts challenging. As the Nobel laureate Albert Einstein said of his Theory of Relativity, "No-one understands relativity; one just gets used to it", or as another Nobel laureate, Richard Feynman, said, "No-one understands quantum mechanics". If people like Einstein and Feynman found the concepts challenging, then the reader is in good company to find the concepts challenging. This book does not 'dumb down' the concepts; it simply neither presents the concepts in hard mathematical terms nor proves the validity of the concepts – you will have to trust your author not to mislead you.

So, if you would like to know what empty space is, what time is, why the universe exists, and why it is as we observe it to be, but you have neither the time nor the confidence to tackle the 'hard mathematics', then this book is for you.

Introduction

To reiterate, you are not getting a 'dumbed down' version of theoretical physics; you are getting a full blown but non-mathematical version.

Chapter 1

An Overview

Modern physics has achieved an understanding of the universe which would be utterly breath-taking to the eyes of people who lived in the stone-age. Even to people like Isaac Newton (1642-1727) or Johannes Kepler (1571-1630) who lived in pre-industrial society, the idea of being able to see billions of light years into outer space or to send probes to distant planets would be dismissed as unbelievably fantastic. Even the most simple of computers or airplanes would astound them. Yet, still physicists are not satisfied. They are not satisfied because they know that things are not quite right

Modern physics is comprised of two separate parts called classical physics and quantum physics. If there is only one universe, there should be only one physics. The two separate parts should fit together somehow.

Classical physics is the special theory of relativity and the general theory of relativity and classical electro-magnetism. Classical physics is the physics we see around us every day. Quantum physics is quantum field theory, QFT. QFT is concerned with atomic particles which are so small, and often so short lived, that no-one has ever seen them. We have seen much evidence for these atomic particles, but, in truth, we have no idea what they really are or any proper understanding of their properties.

An Overview

There is a great philosophical divide between classical physics and quantum physics. Classical physics is deterministic, but quantum physics, at least to our eyes, is based on probability. We can predict with certainty the outcome of a classical physics experiment, but the best we can do with a quantum physics experiment is give a list of the probabilities of possible outcomes. Yet there is determinism within the atomic particles of quantum physics because the probabilities are unchanging; something determines the relative probabilities of possible outcomes.

Physicists believe there is only one universe, and so there should be only one physics. It should be possible to unite both classical physics and quantum physics into a single physics, but efforts over the last century to achieve this unification have completely failed, and no-one understands why these efforts have failed.

Empty space:
There is a part of the universe which is not included within either classical physics or quantum physics. It is a part of the universe which has traditionally, for lack of any idea how to deal with it, been largely ignored by physicists. This ignored part of the universe is empty space and time.

Until recently, no-one had any idea why we live in a 4-dimensional universe or why the 4-dimensional space-time of our universe is of the observed nature (one time dimension and three space dimensions) or how the empty space between groups of galaxies can expand. In short, until recently, non-one had any understanding of empty space and time. However, there have been considerable and unexpected insights into this area of physics recently. Although our understanding is still far from complete, the last few years have seen a revolution in our understanding of empty space.

The revolution in our understanding of empty space has brought with it the prospect of a unification of classical physics and quantum physics and with those prospects a great deal more. It might be that we are now only a few years away from a final theory of all physics, but it is, of course, too early to say. One of the most exciting aspects of these recent new developments is their simplicity.

The simplicity of our present understanding of empty space is two-fold. Firstly, there are no postulates to our understanding of empty space other than the existence of numbers. This contrasts with, say, string theory which must postulate wiggly strings, 10-dimensional space-time, hidden spatial dimensions and other astounding ideas. Secondly, the mathematics is, though cumbersome, very simple; it is so simple, in fact, that a bright 12-year old could be trained to do it in less than a month. The only complications come when we need to tie this simple mathematics into the quite complicated mathematics of quantum physics and classical physics.

The simplicity of our present understanding of empty space allows your author to explain these developments to an intelligent non-physicist, and that is what this book is an attempt to do.

An overview of empty-space:

At first sight, it seems that we live in a universe which is set within four dimensions of space and time and that three of these dimensions are space dimensions and that the other dimension is a time dimension. This is indeed the view that has been taken for millennia.

The space and time of our universe is often viewed as a bare stage upon which various physical objects act out a play. Within this space and time, there seem to be physical objects which interact together as actors on a stage interact together while the stage remains in the background. However, over the previous century, it has become clear that the stage, space-time, itself acts a part and that the actors in some ways might be part of the stage, just different types of empty space.

An understanding of empty space and of the different types of empty space is starting to emerge, and we are starting to approach the view that the physics of our universe is nothing more than interactions between different types of empty space.

We are still a long way from understanding quite how different empty spaces interact and how such interactions would manifest themselves to beings like ourselves in our 4-dimensional space and time. What we observe is not necessarily what happens. Our observations are coloured by the space from which we observe them, by the apparatus we use to make the observations, and by the way our minds interpret what we observe. It is a bit like peering through three sheets of the frosted glass that glaziers use for bathroom windows. In some people's view, it has become pessimistically accepted that the best we can hope to do is to see an average, called an expectation value, of what really happens behind the frosted glass and that we will never know the actual truth.

Your author opines that we can know the actual truth, and that we are rapidly approaching that point. If, as does seem to be the case, the physics of our universe is nothing more than interacting different types of empty space, then once we understand these different types of space, we might be very close to understanding how they interact. Further, once we understand how these empty spaces interact, we might be able to understand how these interactions are perceived on our side of the frosted glass.

The different types of space are just different types of numbers, and it seems that our 4-dimensional space and time is just an average of six copies of a particular type of space (number). We know well all the different types of numbers, and hence the different types of space. There are an infinite number of different types of space, but it seems that only the lesser dimensional spaces have any role in the physics we observe. Perhaps the higher dimensional spaces play a role at ultra-high energies. Once we have understood the physics at low energies, we might want to look at the higher dimensional spaces. We list the important lesser dimensional spaces:

C_1:	1 type of 1-dimensional space	(2.1)
C_2:	2 types of 2-dimensional space	(2.2)
C_3:	4 types of 3-dimensional space	(2.3)
C_4:	24 types of 4-dimensional space	(2.4)

$C_2 \times C_2$:   8 types of commutative 4-dimensional space   (2.5)

$C_2 \times C_2$:   8 types of non-commutative 4-dimensional space   (2.6)

$C_2 \times C_2 \times C_2$:   1024 types of 8-dimensional space   (2.7)

The important spaces to the physicist are the 1-dimensional space, the two 2-dimensional spaces, the eight non-commutative 4-dimensional spaces and, probably, 896 of the 8-dimensional spaces. All the other spaces, most of which we have not listed, seem to play no part in our universe unless they are something to do with dark matter.

The symbols on the left of (2.1) to (2.7) denote finite groups. The different types of numbers (spaces) are within the different finite groups. Finite groups are simple things, and we need only a very shallow understanding of finite group theory to understand the role it plays in the physics of empty space. We will give that very shallow understanding later.

The one 1-dimensional space together with the two 2-dimensional spaces and six of the eight non-commutative $C_2 \times C_2$ spaces, the A_3 spaces, give rise to the whole of classical physics – that is our space-time, gravity and electro-magnetism. It might be that the other two of the eight $C_2 \times C_2$ spaces, the two quaternion spaces, give rise to the weak nuclear force and the 8-dimensional spaces give rise to the strong nuclear force.

So, this book is about different types of empty space. Different types of empty space have different types of rotations in them, and much of this book is concerned with rotations. Associated with different types of rotations are different types of distances and different types of angles. We will spend most of the early chapters of this book discussing the different types of numbers which are the different types of spaces. Toward the end of the book, we will see our 4-dimensional space-time emerge from these different types of spaces (numbers) and bring with it gravity. We will also look at quantum field theory, QFT, and the kind of spaces with which it is concerned.

Chapter 3

We Begin with 1-dimensional Numbers

This chapter is the most mathematical chapter of the book. Once the reader has made it through this chapter, the rest is, mathematically, easy sailing. The maths is no more than the reader has studied at school, but, for some readers, those school days might be a long while ago and their recall of this mathematics might be a little rusty.

Do not be put off by all the equations. You already understand these equations, but you are seeing them in a slightly unfamiliar way. In later chapters, there will be demands made upon the reader of a non-mathematical but conceptual nature. Many of these conceptual demands are challenging; the associated maths is not challenging. Once you get through this chapter, you are, mathematically, home and dry.

Are you sitting comfortably? then we will begin.

We begin:
The reader already understands a lot about numbers. The reader knows the difference between one pint of beer and twenty pints of beer, and so the reader understands the whole numbers like $\{0\}$ and $\{1\}$ and $\{2\}$ and $\{...45,...\}$ and $\{...1,098,765,...\}$. Unless the reader is in the lucky position of having been born into immense wealth, the reader will also

understand negative numbers like $\{-1\}$ and $\{-2\}$ and $\{...,-20,000,...\}$; it's called debt by the bank manager. The reader will also understand fractions; that is numbers like $\left\{\frac{1}{2},\frac{1}{4}\right\}$ and $\left\{\frac{3}{2},\frac{3}{4}\right\}$. Mathematicians have another name for fractions; they call them rational numbers because they are like ratios. The reader will also understand negative fractions because creditors count fractions of whole pounds.

We can, and mathematicians do, arrange these numbers on a line:

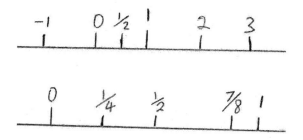

We call this line 'The real number line', and we call the numbers on this line 'The real numbers'. The real number line is infinitely long in both the positive direction and the negative direction because we can continue to count forever.

The real numbers were not always called the real numbers. Prior to circa 1545, the real numbers were simply called numbers, and most people today still simply call them numbers. It was the discovery circa

We Begin with 1-dimensional Numbers

1545 of another kind of numbers that led to mathematicians calling these familiar numbers the real numbers to distinguish them from the newly discovered type of numbers which, in contrast, they called imaginary numbers. We will be looking at the imaginary numbers shortly.

The real number line is always depicted as being straight, but it does not have to be straight. Nor does the distance between two whole numbers always have to be the same, but we always depict the real number line this way.

So far, other than a little nomenclature, you have been taught nothing that you did not already know. Now comes the first important concept.

1-dimensional space:
The real number line is 1-dimensional space. That's it; this is the first important concept.

This is a concept which underpins all of our understanding of our universe. It is not possible to prove that the real number line is 1-dimensional space, but, if we describe 1-dimensional space by listing every property of 1-dimensional space and we describe the real number line by listing every property of the real number line, then we find that the two lists match perfectly[1]. We take this to mean that 1-dimensional space and the real numbers are

---

[1] Your author is hiding something here. No-one has ever seen 1-dimensional space, and so we cannot list the properties of it. All we have seen is our 4-dimensional space-time, and so the best we can do is to assert that 1-dimensional space is the same as 1-dimensional numbers.

the same thing. The equality we assert between the real number line and 1-dimensional space can be questioned; your author does not doubt that the reader is as capable of questioning this assertion as any great philosopher. The upshot is that our present understanding of the universe assumes that the real number line is 1-dimensional space.

Does the real number line have gaps in it? Looking at the diagram above, we see that there is a gap between $\frac{1}{2}$ and $\frac{1}{4}$, but we know there is the number $\frac{3}{8}$ between these two numbers. What about the gap between $\frac{1}{2}$ and $\frac{3}{8}$? The number $\frac{7}{16}$ is between $\frac{1}{2}$ and $\frac{3}{8}$. With a little thought, the reader will see two things:

- a) No matter how close two fractions are to each other, we can always find a fraction between those two fractions.
- b) There are always gaps between two fractions. The gaps might be extremely small, but they are not zero.

At this point, we might have to conclude that the real number line has gaps in it. This is not how we envisage 1-dimensional space. But wait, there are some other kinds of real numbers which we have not considered. There are a kind of real numbers which mathematicians call irrational numbers; these are numbers which cannot be written as a fraction. This means that irrational numbers must be numbers which do not have a final digit. If a number has a final digit, we can write it as a fraction by simply putting

We Begin with 1-dimensional Numbers

it over a 1 followed by the requisite number of zeros. For example:

$$0.12345 \to \frac{12345}{100,000} \qquad (3.1)$$

An example of an irrational number is the number:

$$0.12345678910111213141516 17... \qquad (3.2)$$

This number is formed by simply adding the next whole real number on to the end. This number has no final digit. Other examples are the square roots of numbers which end in 2.

Assume there is a number, called a square root, with a final digit, and, assume that when this square root is multiplied by itself, it produces a number which ends in 2. Since that square root has a final digit, that square root must end in one of ten digits, either $\{0,1,2,3,4,5,6,7,8,9\}$. If we multiply together two numbers which end in 0 we will produce a number which ends in 0. Similarly:

$$\begin{aligned} ....0 \times ....0 &= ....0 \\ ....1 \times ....1 &= ....1 \\ ....2 \times ....2 &= ....4 \\ ....3 \times ....3 &= ....9 \\ ....4 \times ....4 &= ....6 \end{aligned} \qquad (3.3)$$

$$....5 \times ....5 = ...5 \qquad (3.4)$$

$$.....6 \times ....6 = ...6$$
$$.....7 \times ....7 = ...9$$
$$.....8 \times ....8 = ...4 \quad (3.5)$$
$$.....9 \times ....9 = ...1$$

We see that there are no square numbers which end in 2 (or 3 & 7 & 8), and so we know that the square roots of such numbers are irrational[2].

Although the demonstration of it is beyond this book, there are an uncountable number of irrational numbers between any two fractions. This means that the real number line is continuous, which is the same as space. The reader is directed to the many mathematical texts upon irrational numbers if they wish to delve deeper into these numbers.

The concept which the reader needs to take from the previous section is no more than the fact that 1-dimensional space is the same thing as 1-dimensional numbers – that is the real number line.

A taste of what is to come:

If 1-dimensional space is the same as 1-dimensional numbers, we might expect that 2-dimensional space is the same as 2-dimensional numbers, and so we might expect to discover a type of 2-dimensional numbers; our expectation would be correct. In fact, there are two types of 2-dimensional number and two types of 2-dimensional space to match them. We

---

[2] We have shown the above in number-base ten. The number-base we use is arbitrary. If a number is irrational in one number-base, then it is irrational in all number-bases.

We Begin with 1-dimensional Numbers

might expect that our 4-dimensional space-time is the same as 4-dimensional numbers; our expectation would be incorrect. Our 4-dimensional space-time is something different from a type of numbers.

What are numbers?

Technically, a set of numbers, like the real numbers, is a set of mathematical objects that have thirteen properties[3]. Some of these properties are quite technical, but others are quite simple. We list a few of the simple properties:

a) A number multiplied by a number produces a number.
b) If two numbers multiplied together equal zero, then at least one of the numbers must be zero.
c) Corresponding to every number, there is an inverse number such that these two numbers multiplied together equal one. For example, $\frac{1}{2} \times 2 = 1$.

The first of these properties is called multiplicative closure. It is basically that a donkey mated with a donkey must produce a donkey. There are things in mathematics which do not have multiplicative closure. A 1×4 matrix multiplied by a 4×1 matrix produces a 1×1 matrix, and two vectors 'dotted' together produce a number and not a vector; mating two donkeys has produced a duck.

---

[3] The set satisfies the thirteen axioms that define a division algebra.

The second of these properties is called absence of zero divisors. You cannot choose two non-zero real numbers and multiply them together to make zero. Absence of zero divisors might seem stark raving obvious, and it is with real numbers, but there are things in mathematics which do not have absence of zero divisors. In general matrices do not have this property and there are individual elements of Clifford algebras which, although themselves non-zero, when multiplied together produce zero.

The third of these properties is called the existence of multiplicative inverses. Again, the existence of multiplicative inverses is stark raving obvious within the real numbers, but there are matrices and elements of Clifford algebras which do not have multiplicative inverses.

In short, a set of numbers is some set of mathematical objects which has the same thirteen defining properties as have the real numbers.

Mathematicians call such a set that has the thirteen defining properties of the real numbers a division algebra. The real numbers are a 1-dimensional division algebra.

Important Note: Within the real numbers, we have that the order of multiplication does not affect the result of the multiplication. We have $2 \times 3 = 3 \times 2$. This is called multiplicative commutativity.

Multiplicative commutativity is not one of the thirteen properties that define a type of numbers, and there are types of numbers within which the order of the multiplication does matter – we will meet them later.

We Begin with 1-dimensional Numbers

Well, we said we were not going to do any hard mathematics, and yet here we are writing in quite technical terms. Yes the terms do sound technical, but I am quite sure the reader understands the concepts behind them. There is nothing difficult about realising that multiplying, say, 5 and 6 together will produce a number. Nor is there anything difficult about realising, say, 5 multiplied by 6 is not zero or that $5 \times \frac{1}{5} = 1$ and $\sqrt{2} \times \frac{1}{\sqrt{2}} = 1$.[4] It really is that simple. Of course, if you want to sound really posh, you can learn the technical terminology.

To sum up: a set of mathematical objects is a type of numbers if they have the thirteen defining properties of the real numbers.

---

[4] The $\sqrt{}$ sign means square root. We have $\sqrt{4} = \pm 2$.

Chapter 4

2-dimensional numbers:

We all know what 2-dimensional space is. It's that flat sheet of paper on your desk. So what would a 2-dimensional number look like? For a start, it will have to be two 1-dimensional numbers, one for each dimension; that is two real numbers. In just the same way that the 1-dimensional numbers are positions in 1-dimensional space, that is positions on the real number line, so it is that 2-dimensional numbers are positions on the flat sheet of paper. You can think of a 2-dimensional number as being comprised of the latitude and longitude of its position in the flat sheet of paper. Of course, the flat sheet of paper is infinitely extensive in both directions as the real number line is infinitely long.

The 2-dimensional numbers are sometimes written as a pair of real numbers, $(2,3)$. Like latitude and longitude, we have to be careful to keep these in order, and so we often attach a letter i to the second number to identify it; we then write the 2-dimensional number as $2+3i$. The plus sign is questionable notation. It is much better notation to write our 2-dimensional numbers as square matrices. A matrix is just a box of numbers:

$$2+3i \equiv \begin{bmatrix} 2 & 3 \\ -3 & 2 \end{bmatrix} \tag{4.1}$$

We generalise this to:

$$\begin{bmatrix} a & b \\ -b & a \end{bmatrix} \quad (4.2)$$

In which both $\{a,b\}$ are just real numbers like 2 or 3. The reader will have to take the author's word for how we multiply matrices together, but we have:

$$\begin{bmatrix} a & b \\ -b & a \end{bmatrix}\begin{bmatrix} c & d \\ -d & c \end{bmatrix} = \begin{bmatrix} ac-bd & ad+bc \\ -(ad+bc) & ac-bd \end{bmatrix} \quad (4.3)$$

Notice how the form of the matrix is maintained through multiplication. This is a duck mated with a duck producing a duck; this is called multiplicative closure to be technical. Although we are not going to do it, it is quite easy to show that the above form of matrix, (4.2), has all the defining properties that the real numbers have; that is, it satisfies the axioms of a division algebra. In other words, the matrix (4.2) is a type of number. It is a type of number because it has the thirteen defining properties that numbers have. It is a 2-dimensional type of number, and it corresponds to the 2-dimensional flat sheet of paper on your desk. We call this type of 2-dimensional number the Euclidean complex numbers or just complex numbers for brevity. Within mathematics, the Euclidean complex numbers are denoted by the symbol $\mathbb{C}$.

Determinants and distance functions:
Every square matrix has a single real number called 'the determinant of the matrix' associated with it. The determinant of a matrix is the 'length' of the

matrix raised to some power. By 'length' we mean the distance that the matrix is from the same size matrix that is all zeros. There is a definite way of calculating the determinant of a matrix. The determinant of the matrices of the form of (4.2), that is the Euclidean complex numbers, is the origin of the Pythagoras theorem:

$$\det = a^2 + b^2$$
$$d^2 = a^2 + b^2 \tag{4.4}$$

We have use the variable d to mean distance in the immediately above (4.4). The reader does not need to know how a determinant is calculated but does need to know that the determinant gives the 'length of the number' – the distance from zero, the origin, to the number. In general, when dealing with square matrices which are a type of number, the determinant is the distance function of that type of number.

Aside: We can write the real numbers as 1×1 matrices. In this case, the determinant of the 1-dimensional number $[a]$ is just a. This is how far a is from the number zero.

A little history:

These 2-dimensional numbers were discovered by the Italian mathematician Gerolamo Cardano (1501-1576). He published his discovery in his book *Ars Magna* in 1545. Also influential in the discovery were Niccolo Foantana Tartaglia and Cardano's student Lodovico Ferrari.

The discovery of this previously unknown type of numbers shocked the academic world at the time. No-one had ever thought it possible that there could be more than one type of numbers, and the acceptance of Cardano's discovery was reluctant. An expression of this reluctance is that the newly discovered type of numbers were called imaginary numbers as if they did not really exist. Today, we call them Euclidean complex numbers but this is a very recent change. Your author who, at least in his heart is still a young man, recalls these numbers being called imaginary numbers when he first studied them in the 1970's. There are still many books available that refer to these 2-dimensional numbers as imaginary numbers. Even today, at the heights of theoretical physics, there is still a reluctance to accept these Euclidean complex numbers. Your author recalls a conversation with a leading researcher into quantum field theory in which this leading researcher said, "The most mysterious thing to me about quantum physics is than we have to use the imaginary numbers to do it".

Aside: Cardano had a difficult life. He was born illegitimate and lived in poverty all his life. In 1560, he had to watch his son being executed for murder.

Cardano did much more than only mathematics. His achievements include the invention of the combination lock and the invention of the universal joint still used in motor vehicles today.

In those days, the 2-dimensional numbers were associated with solving quadratic equations rather than with 2-dimensional space. It was the English mathematician Wallis (1616-1703) who in 1673 first associated the 2-dimensional numbers with 2-

dimensional space. The association was made again by Wessel (1745-1818) in 1798, by Gauss (1777-1855) in 1799, and by Argand (1768-1822) in 1806. Argand gets the credit for it. The equivalent of the 1-dimensional real number line for the 2-dimensional Euclidean complex numbers is just a flat sheet of paper with axes on it; it is called the Argand diagram.

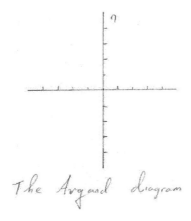

The Argand diagram

Rotation:
We cannot rotate in 1-dimensional space. We can rotate in 2-dimensional space. There is no concept of rotation within the 1-dimensional real numbers. There is rotation within the 2-dimensional Euclidean complex numbers.

We can specify any point in the 2-dimensional plane by two numbers which are the distance along the horizontal axis and the distance along the vertical axis. This is called using Cartesian co-ordinates.

2-dimensional Numbers

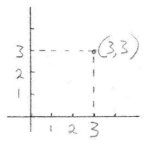

We can also specify the same point as two numbers which are the angle from the horizontal axis and the radial distance from the origin at $(0,0)$. This is called using polar co-ordinates. We can write this as (r,θ) where r is the distance from the origin to the point and θ is the angle from the horizontal axis.

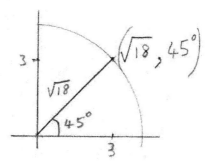

The radial distance is $\sqrt{18}$ because, using the Pythagoras theorem, we have:

$$d = \sqrt{3^2 + 3^2} = \sqrt{18} \qquad (4.5)$$

A little trigonometry:
The reader will recall using trigonometry at school to calculate the lengths of sides of right-angled triangles

or the heights of distant church steeples. The reader will have heard of the trigonometric functions known as the sine function, sin( ), and the cosine function, cos( ), and many other trigonometric functions like the tangent function. Unfortunately, it is likely that trigonometry was never explained properly to the reader and the reader might think of it as being difficult. Lucky for us, we do not need the difficult bits. All we need is to know about are the cosine function and the sine function.

The cosine of a given angle, say 45^0, written as $\cos(45^0)$, is just the distance along the horizontal axis of a point that is radial distance of one unit from the origin and at 45^0 to the horizontal axis. The sine of a given angle, say 45^0, written as $\sin(45^0)$, is just the distance up the vertical axis of a point that is radial distance of one unit from the origin and at 45^0 to the horizontal axis.

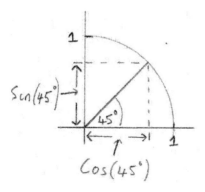

That's it, and you thought trigonometry was hard. Using the Pythagoras theorem, we find:

2-dimensional Numbers

$$\cos^2\theta + \sin^2\theta = 1 \qquad (4.6)$$

Suppose the angle is 0^0 and the radial line is thus lying along the horizontal axis. Clearly, the distance along the horizontal axis to a point on the horizontal axis which is one unit from the origin is one unit. We have $\cos(0^0) = 1$. However, the height up the vertical axis of a point on the horizontal axis is zero, and so we have $\sin(0^0) = 0$. Suppose the angle is 90^0 and the radial line is thus lying along the vertical axis. Clearly, the distance along the vertical axis to a point on the vertical axis which is one unit from the origin is one unit, and we have $\sin(90^0) = 1$. However, the distance along the horizontal axis of a point on the vertical axis is zero, and we have $\cos(90^0) = 0$. We can draw a graph of how the cosine function varies as the radial line rotates:

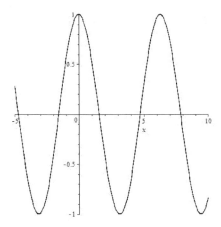

We have a wave. The sine function is the same but is 90^0 in front of (out of phase with) the cosine function.

There is something of interest here. Within physics, we find waves everywhere and the mathematical description of those waves involves the cosine function and the sine function. The cosine function and the sine function exist in only the Euclidean complex number space, and there are no other functions which wave at you. Therefore, the wave phenomena of the physical world must happen within the 2-dimensional space which is the Euclidean complex numbers. We need to modify this a little. In due course, we will see wavy trigonometric functions within the 4-dimensional quaternion space. However, the quaternion trigonometric functions are never used to describe waves in the physical world.

The polar form and rotation matrices:
Using the trigonometric functions, we can specify any point in the 2-dimensional plane as a rotation and a radial length. This is writing the complex number in polar form. So instead of writing a 2-dimensional Euclidean complex number as (4.2), we can write a 2-dimensional Euclidean complex number as:

$$(r,\theta) \equiv \begin{bmatrix} r & 0 \\ 0 & r \end{bmatrix} \begin{bmatrix} \cos\theta & \sin\theta \\ -\sin\theta & \cos\theta \end{bmatrix} \quad (4.7)$$

The matrix with the two r's is just a real number written in 2×2 matrix form; it is the radial distance of the complex number from the origin. The matrix with the $\theta's$ is a rotation matrix. Consider the 2-dimensional Euclidean complex number:

$$(4,3) \equiv \begin{bmatrix} 4 & 3 \\ -3 & 4 \end{bmatrix} \qquad (4.8)$$

Using the Pythagoras theorem:

$$d = \sqrt{3^2 + 4^2} = \sqrt{9+16} = \sqrt{25} = 5 \qquad (4.9)$$

We see that this complex number is distance 5 units from the origin. If we multiply this complex number by the rotation matrix in (4.7), we get another complex number:

$$\begin{bmatrix} \cos\theta & \sin\theta \\ -\sin\theta & \cos\theta \end{bmatrix} \begin{bmatrix} 4 & 3 \\ -3 & 4 \end{bmatrix}$$
$$= \begin{bmatrix} 4\cos\theta - 3\sin\theta & 3\cos\theta + 4\sin\theta \\ -(3\cos\theta + 4\sin\theta) & 4\cos\theta - 3\sin\theta \end{bmatrix} \qquad (4.10)$$

Using the Pythagoras theorem on this other complex number gives the distance from the origin to be:

$$\begin{aligned} d &= \sqrt{(4\cos\theta - 3\sin\theta)^2 + (3\cos\theta + 4\sin\theta)^2} \\ &= \sqrt{\begin{aligned} &16\cos^2\theta + 9\sin^2\theta - 24\cos\theta\sin\theta \\ &+ 9\cos^2\theta + 16\sin^2\theta + 24\cos\theta\sin\theta \end{aligned}} \\ &= \sqrt{25(\cos^2\theta + \sin^2\theta)} \\ &= 5 \end{aligned} \qquad (4.11)$$

We see that, because $\cos^2\theta + \sin^2\theta = 1$, for any angle, θ, the distance from the origin is unchanged when we multiply by the rotation matrix. That is rotation.

Though-out the whole of engineering, physics and mathematics the rotation matrix in (4.7) is recognised as being a 2-dimensional rotation, yet there is something very strange about it that escaped the notice of mathematicians for 150 years after it was first written down. That 'something strange' is very closely connected to quantum mechanics and the intrinsic spin of an electron.

The 'something strange' is that rotation within the 2-dimensional Euclidean complex numbers is not rotation about an axis. This is a 2-dimensional rotation in a 2-dimensional space. There is no third dimension sticking out of the 2-dimensional plane to be an axis of rotation. The intrinsic spin of sub-atomic particles like the electron is rotation without an axis.

Summary of this chapter:
    a) The numbers we learnt about at infant school are called the real numbers.
    b) The real numbers are the same thing as 1-dimensional space.
    c) Any set of mathematical objects that have the thirteen defining properties that define the real numbers is also a type of numbers.
    d) There are 2-dimensional numbers which match 2-dimensional space. They are called Euclidean complex numbers and are written as:

$$\begin{bmatrix} a & b \\ -b & a \end{bmatrix} \quad (4.12)$$

or as:

$$\begin{bmatrix} r & 0 \\ 0 & r \end{bmatrix} \begin{bmatrix} \cos\theta & \sin\theta \\ -\sin\theta & \cos\theta \end{bmatrix} \quad (4.13)$$

e) The determinant of the matrix, (4.12), gives the 'length' of the number (the distance from zero).
f) We can rotate in these 2-dimensional numbers just as we can rotate in 2-dimensional space. Rotation is done with a rotation matrix of the form:

$$\begin{bmatrix} \cos\theta & \sin\theta \\ -\sin\theta & \cos\theta \end{bmatrix} \quad (4.14)$$

g) The functions in this rotation matrix are the trigonometric functions of 2-dimensional Euclidean space. They are projections on to the axes of the space from a point on the unit circle in the space.

Chapter 5

Fibre Bundles

There is a fundamental difference between the way Isaac Newton viewed space and time and the way that we view space and time today. Newton thought of time as a 1-dimensional space completely separate from the 3-dimensional space we see around us. He took the view that there was a copy of our 3-dimensional space fixed to every point of the 1-dimensional space that is time. Newton viewed space and time as what mathematicians call a fibre bundle.

Because we are artistically disadvantaged, we illustrate Newton's view with a 2-dimensional plane instead of the 3-dimensional space:

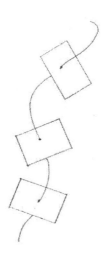

Fibre Bundles

A fibre bundle is a few (could be one or many) separate types of space fixed on to an underlying space at every point of that underlying space. There is nothing that says the two or more spaces that are tied together in a fibre bundle have to be of the same dimension or of any particular dimension.

Our modern view is that 3-dimensional space and 1-dimensional time are not two separate spaces but together form a single 4-dimensional space called 4-dimensional space-time. This only one space-time view is a fundamentally different view of space and time from Newton's view of space and time as two separate spaces.

We have around us three spatial dimensions and one time dimension. The modern view is that these four dimensions form a single 4-dimensional space-time. This means that, since the four dimensions are in the same space, we can rotate in a space-time plane as well as rotating in a purely spatial plane. This ability to rotate in any plane formed of any two axes, including the time axis, is exactly what we mean by saying our 4-dimensional space-time is a single space rather than two separate spaces tied together as a fibre bundle. We cannot form a 2-dimensional plane from two axes that are in separate spaces.

The Pythagoras theorem (distance function) of our 4-dimensional space-time is:

$$d = \sqrt{t^2 - x^2 - y^2 - z^2} \qquad (5.1)$$

This is not the distance function of any type of 4-dimensional number, and so we know that the 4-dimensional space-time in which we sit is not a type of numbers. We now realise that we have two types of empty space; one type is just a type of numbers, and the other type is not a type of numbers. In due course, we will see how our 4-dimensional space-time emerges as an 'average' of six spaces which are types of numbers.

We will meet fibre bundles later in this book. Modern quantum field theory, QFT, holds that the physical forces other than gravity arise from different types of space being fixed to our underlying space-time as a fibre bundle. Those different types of space are different types of numbers, and one of them is the 2-dimensional Euclidean complex numbers we met previously. Modern parlance uses the name $U(1)$ rather than Euclidean complex numbers and thinks of rotation in the Euclidean complex plane as a Lie group, but that is getting unnecessarily technical.

Chapter 6

Strange Rotations

Spinor rotations:
We can rotate within the three spatial dimensions of our 4-dimensional space-time. Imagine a spinning metal disc. However we spin the metal disc, there is a line perpendicular to the plane of the disc and through the centre of the disc which we call the axis of rotation of the disc.

A little less than 300 years ago, the mathematician Leonard Euler (1707-1783) proved that all rotations within the 3-dimensional spatial part of our 4-dimensional space-time are 2-dimensional rotations about an axis. Every rotation the reader has ever seen is a rotation about an axis. How can we have a rotation that is not about an axis?

Approximately 100 years ago, physicists discovered sub-atomic particles called electrons. Electrons rotate in two ways. The first way is that they orbit an atomic nucleus. There is a spatial axis associated with this orbital rotation which is perpendicular to the orbital plane, just as we would expect it to be. The second way that electrons rotate is entirely different in that it is not rotation about an axis. (It might not even be 2-dimensional rotation, but we'll come to that later.) This 'not rotate about an axis' is called spinor rotation by physicists.

Let us consider the rotation matrix we met earlier, (4.14). This is a 2-dimensional rotation in a 2-

dimensional space. There is no third spatial direction to be an axis. This is 2-dimensional rotation without an axis. Yes! The concepts are challenging, but you completely understand that 2-dimensional space has no third dimension. The maths is not hard; it's just the concept. Do not be dismayed, neither your author nor any mathematician or physicist understands the concept any better than you do[5].

Well, that is spinor rotation; it is not rotation about an axis. In quantum physics, spinor rotation is known as intrinsic spin. Spinor rotation is the type of rotation we get in every space which is a type of numbers. We have rotation about an axis in only our 4-dimensional space-time.

There are some other peculiarities about spinor rotation. The trigonometric functions of the n-dimensional spinor rotation matrix accept $(n-1)$ angles rather than only one angle. In 2-dimensional space, $(n-1)=1$. Some spinor rotations are also what physicists call double cover. There is more of all this later.

Space-time rotations:
So far, we have considered 2-dimensional rotation in spaces comprised of only spatial axes. There is time in our universe. Time is a dimension of our 4-dimensional space-time. Can we rotate 2-dimensionally in a space comprised of one spatial axis and one time axis? Isaac Newton would have

---

[5] Technically, the rotation matrix has no eigenvalues and associated eigenvectors which are independent of the angle variable.

said no. Albert Einstein said yes. We can do this because our 4-dimensional space-time is a single space rather than a fibre bundle.

As we rotate, the ratio of the lengths of the horizontal and vertical components of our position change:

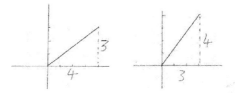

In spatial rotation, these two lengths are both spatial and their ratio is a gradient. If one of the axes is time and the other is space, their ratio is a ratio of an amount of time to an amount of length. A ratio of length to time is a velocity like a mile per hour or twenty metres per second. Rotation in 2-dimensional space-time is a change in the ratio of length to time and this is a change of velocity; it is like changing from ten miles per second to twenty miles per half-second (forty miles per second).

So, when you put your foot on the accelerator pedal and increase your velocity, what you are really doing is rotating in 2-dimensional space-time. The maths is simple; it's the concept that gives pause for thought.

Of course, where there is a 2-dimensional rotation, there is a 2-dimensional rotation matrix; that's a 2×2 box of numbers. Where there is a 2-dimensional rotation matrix, there is a 2-dimensional type of numbers. Just as we found a Euclidean type of 2-dimensional complex numbers corresponding to 2-dimensional Euclidean space (the flat sheet of paper), there should be a type of 2-dimensional space-time numbers corresponding to the 2-dimensional space-time plane. There is; they are called the hyperbolic complex numbers. They were first discovered by James Cockle (1819-1895) in 1848. They are very closely related to the complex numbers we met above[6]. We have:

$$\exp\left(\begin{bmatrix} a & b \\ -b & a \end{bmatrix}\right) = \begin{bmatrix} r & 0 \\ 0 & r \end{bmatrix}\begin{bmatrix} \cos b & \sin b \\ -\sin b & \cos b \end{bmatrix}$$

$$\exp\left(\begin{bmatrix} a & b \\ b & a \end{bmatrix}\right) = \begin{bmatrix} r & 0 \\ 0 & r \end{bmatrix}\begin{bmatrix} \cosh b & \sinh b \\ \sinh b & \cosh b \end{bmatrix}$$

(6.1)

The $\cosh(\ )$ function and $\sinh(\ )$ function are the trigonometric functions of 2-dimensional space-time, and the matrix in which they appear is the rotation matrix of 2-dimensional space-time. The $\exp(\ )$ is an enormously important function which appears all over the place in mathematics and physics; here, we are using it to convert the Cartesian form of a type of 2-dimensional number into its polar form.[7]

---

[6] They both derive from the finite group C_2. They are as brother and sister.

[7] For technical reasons, the Cartesian form of the hyperbolic complex numbers is not a division algebra but the polar form is a division algebra. This failing corresponds to the fact that we cannot move faster than the speed of light or travel backwards in time.

The trigonometric functions can be written as infinite sums; infinite sums are called series by mathematicians. We have:

$$\cosh x = 1 + \frac{x^2}{2!} + \frac{x^4}{4!} + \frac{x^6}{6!} + \ldots$$
$$\cos x = 1 - \frac{x^2}{2!} + \frac{x^4}{4!} - \frac{x^6}{6!} + \ldots$$
(6.2)

Only a few minus signs different. And:

$$\sinh x = \frac{x^1}{1!} + \frac{x^3}{3!} + \frac{x^5}{5!} + \ldots$$
$$\sin x = \frac{x^1}{1!} - \frac{x^3}{3!} + \frac{x^5}{5!} - \ldots$$
(6.3)

We see that the two types of 2-dimensional trigonometric functions are very closely related to each other[8].

In general, trigonometric functions are very closely related to the exponential function which is:

$$\exp x = 1 + \frac{x^1}{1!} + \frac{x^2}{2!} + \frac{x^3}{3!} + \frac{x^4}{4!} + \frac{x^5}{5!} + \ldots \quad (6.4)$$

Distance in 2-dimensional space-time:
The determinant of the Cartesian form of the hyperbolic complex numbers gives the distance

---

[8] Calculators use these series to evaluate the trigonometric function of a given angle.

function (Pythagoras theorem) for 2-dimensional space-time:

$$d = \sqrt{t^2 - z^2} \qquad (6.5)$$

Notice the minus sign. Within mathematics, the hyperbolic complex numbers are denoted by the symbol $\mathbb{S}$.

Although your author will explain it in more detail later, the special theory of relativity is the fact that engines and chemical reactions and all physical phenomena work exactly the same regardless of the direction in which the apparati are pointing in 2-dimensional space-time. It is just like a car engine works the same when the car is heading north as it does when the car is heading east. So, an electrical circuit works the same when it is on a rocket travelling at 100,000 miles per second as it does when it is standing still. The only difference is velocity, and a change of velocity is just a change of direction in space-time.

Well the concept might have caused the reader to pause, but the idea of rotating in 2-dimensional space-time is quite straight-forward.

What about higher dimensional rotations?

Above, (6.1), we have seen 2-dimensional numbers put into their polar form by the exponential function. We have seen the concept of 2-dimensional rotation emerge from within the 2-dimensional numbers. The reader might have heard of the 4-dimensional numbers called quaternions. We have the exponential of a quaternion:

$$\exp\left(\begin{bmatrix} a & b & c & d \\ -b & a & -d & c \\ -c & d & a & -b \\ -d & -c & b & a \end{bmatrix}\right) \tag{6.6}$$

This is going to blow the reader's mind. Before we show this, we will prepare the reader a little.

We could write a quaternion as four real numbers, (a,b,c,d), but we have to keep track of which position each number is in, and so, like the complex numbers, quaternions are often written as:

$$a + ib + jc + kd \tag{6.7}$$

We will write quaternions as a matrix in (6.6).

The quaternions are one of several types of 4-dimensional numbers which are not multiplicatively commutative. The reader will perhaps remember that multiplicative commutativity is not one of the thirteen defining properties of a type of numbers. We must explain what these terms mean.

We are used to the idea that the order of multiplication of two numbers does not matter; we get the same answer regardless of the order in which the numbers are written. Within the real numbers, we have 4×5 is the same as 5×4. This is called multiplicative commutativity. This is a property of the 1-dimensional real numbers and of both types of 2-dimensional numbers. It is also a property of all four types of 3-dimensional numbers and of most types of 4-dimensional numbers. However, there are eight types of 4-dimensional numbers which have all the defining properties of the real numbers but they are not multiplicatively commutative numbers. The quaternions are one of these eight types of numbers[9].

Within the quaternions, and all non-commutative types of numbers, the order of multiplication does matter; we get different answers depending on the order in which we do the multiplication. In the case of the quaternions, we have:

$$(a+ib+jc+kd)(w+ix+jy+kz) \neq \\ (w+ix+jy+kz)(a+ib+jc+kd) \quad (6.8)$$

The quaternions are considered to be numbers because they have so much in common with the real numbers. They have the thirteen defining properties of numbers in common with the real numbers. Technically, the quaternions are a non-commutative division algebra whereas the real numbers are a commutative division algebra.

The quaternions were discovered by the Irish mathematician William Rowan Hamilton (1805-1865). His discovery was part of a search for 3-dimensional numbers and for higher dimensional numbers in general. The search had started shortly after the discovery of the Euclidean complex numbers by Cardano in 1545 but had been done without success before Hamilton's discovery of the quaternions in 1863.

---

[9] The other seven types are six A_3 types and the anti-quaternions.

Within mathematics, the quaternion numbers are denoted by the symbol $\mathbb{H}$.

In the same way that each of the 2-dimensional types of complex numbers have a rotation matrix, the quaternions have a rotation matrix. It is:

$$\mathbb{H}_{Rot} = \begin{bmatrix} \cos(\lambda) & \frac{b}{\lambda}\sin(\lambda) & \frac{c}{\lambda}\sin(\lambda) & \frac{d}{\lambda}\sin(\lambda) \\ -\frac{b}{\lambda}\sin(\lambda) & \cos(\lambda) & -\frac{d}{\lambda}\sin(\lambda) & \frac{c}{\lambda}\sin(\lambda) \\ -\frac{c}{\lambda}\sin(\lambda) & \frac{d}{\lambda}\sin(\lambda) & \cos(\lambda) & -\frac{b}{\lambda}\sin(\lambda) \\ -\frac{d}{\lambda}\sin(\lambda) & -\frac{c}{\lambda}\sin(\lambda) & \frac{b}{\lambda}\sin(\lambda) & \cos(\lambda) \end{bmatrix}$$

$$\lambda = \sqrt{b^2 + c^2 + d^2}$$

(6.9)

The quaternion trigonometric functions in this rotation matrix are very closely related to the sine and cosine functions of the 2-dimensional Euclidean complex numbers. They differ in only the coefficients $\frac{b}{\lambda}$ and that they accept three real numbers as angles under a square root sign rather than only one real number as an angle.

This is a 4-dimensional rotation. It is a spinor rotation. It is a 4-dimensional rotation in 4-dimensional space; it has no axis of rotation. No human has ever seen anything like this in the physical world. In our space-time, we have only two dimensional rotations. In the quaternion rotation matrix, the trigonometric functions accept three real numbers as a 4-dimensional angle (under a square root sign). The 2-dimensional trigonometric functions to which we are accustomed accept only one real number as an angle. So, an angle in quaternion space is three real numbers[10]. This is a general phenomenon in spinor spaces. The 3-dimensional spinor spaces have angles which are two real numbers; the 7-dimensional spinor spaces have angles which are six real numbers etc..

To explain quaternion rotation more easily, we will set two of the variables in the quaternion angle to zero. This is no more than selecting a particular co-ordinate system; we are not reducing this to a 2-dimensional rotation. We have:

$$\mathbb{H}_{Rot} = \begin{bmatrix} \cos(\lambda) & \dfrac{b}{\lambda}\sin(\lambda) & 0 & 0 \\ -\dfrac{b}{\lambda}\sin(\lambda) & \cos(\lambda) & 0 & 0 \\ 0 & 0 & \cos(\lambda) & -\dfrac{b}{\lambda}\sin(\lambda) \\ 0 & 0 & \dfrac{b}{\lambda}\sin(\lambda) & \cos(\lambda) \end{bmatrix}$$

$$\lambda = \sqrt{b^2}$$

(6.10)

Now, we can take the square root of any real number to be either positive of negative. We have:

---

[10] Mathematicians call this 3-tuple of real numbers a vector.

$$(-a) \times (-a) = a^2$$
$$(a) \times (a) = a^2 \tag{6.11}$$

You learnt this at school. With (6.10) in mind, this means that we get rotation in both the clockwise direction and the anti-clockwise direction at the same time. Another way to see this is to note that the position of the minus sign before the sin( ) function in the 2×2 blocks of complex numbers dictates whether the rotation is clockwise or anti-clockwise. Within the quaternion rotation matrix, we have the minus sign in both positions. I told you the concepts were challenging. This is what mathematicians call double cover. The quaternion rotation is the same as a rotation in the Euclidean complex numbers – that's the kind of rotation to which we are accustomed – except that we have rotation in both directions together rather than in only one direction. If you think about it, it is more symmetrical to rotate in both directions at the same time than to rotate in only one direction. Perhaps it is time for the pub.

It gets even more astonishing. If it wasn't time for the pub before, it soon will be. We rotate a point in space by multiplying by a rotation matrix. The reader saw this above with the Euclidean complex numbers, (4.10). We can multiply with the rotation matrix on the left or with the rotation matrix on the right. With the commutative 2-dimensional complex numbers, the position of the rotation matrix makes no difference:

$$\begin{bmatrix} \cos\theta & \sin\theta \\ -\sin\theta & \cos\theta \end{bmatrix} \begin{bmatrix} 4 & 3 \\ -3 & 4 \end{bmatrix}$$
$$= \begin{bmatrix} 4\cos\theta - 3\sin\theta & 3\cos\theta + 4\sin\theta \\ -(3\cos\theta + 4\sin\theta) & 4\cos\theta - 3\sin\theta \end{bmatrix} \quad (6.12)$$

$$\begin{bmatrix} 4 & 3 \\ -3 & 4 \end{bmatrix} \begin{bmatrix} \cos\theta & \sin\theta \\ -\sin\theta & \cos\theta \end{bmatrix}$$
$$= \begin{bmatrix} 4\cos\theta - 3\sin\theta & 3\cos\theta + 4\sin\theta \\ -(3\cos\theta + 4\sin\theta) & 4\cos\theta - 3\sin\theta \end{bmatrix} \quad (6.13)$$

So, within the Euclidean complex numbers, rotation on the left through a given angle and rotation on the right through the same given angle both move a point to the same new point – same angle, same rotation – that appeals to our common sense.

Because the quaternions are non-commutative, rotating through a given angle by multiplying by the quaternion rotation matrix on the left produces a different answer to rotating through the same angle by multiplying by the quaternion rotation matrix on the right. We have rotation in a given direction through a given angle moves a point to two different places. Your author is used to this stuff, and your author still thinks that's weird, but the maths is simple and straight-forward. We have:

$$\mathbb{H}_{Rot} Q \neq Q \mathbb{H}_{Rot} \quad (6.14)$$

For many decades, mathematicians, actually mathematicians in a specialist area of mathematics called Clifford algebra, did not like this 'one angle but two different rotations'. They found a way to circumvent their dislike; multiply by the rotation matrix on both the left and the right at the same time. Thus, Clifford algebraists would write a quaternion rotation as:

$$\mathbb{H}_{Rot} \mathcal{Q} \mathbb{H}_{Rot} \qquad (6.15)$$

This does work; we have a single rotation for a single angle, but it is a distortion of the mathematics made only to suit human prejudice, and there is a fly in the ointment.

Now let us arrange our co-ordinate system to give us a seeming 2-dimensional rotation like (6.10). Now let us rotate through the angle 360^0. We know that the sine and the cosine functions repeat every 360^0, (see above graph of the cosine function) and so this is a full rotation. We put $b = 360^0$ and we find that our quaternion has rotated through 720^0. That is 360^0 for each $\mathbb{H}_{Rot}$. The quaternion rotates twice as much as it should do. Of course this is an illusion, the quaternion rotates through twice the angle it should do because human beings attached two rotation matrices to the quaternion rather than one. Even so, the reader might come across literature which describes the electron as rotating through 720^0 before it gets back to where it started. Electrons are associated with non-commutative types of numbers.

The reader might see this double rotation, (6.15), presented as:

$$X^{-1}\vec{V}X \qquad (6.16)$$

and called double cover rotation.

Stranger and stranger:
Within our 4-dimensional space-time, there are six ways we can select two variables from the four variables $\{t, x, y, z\}$ which are the dimensions of our 4-dimensional space-time. That is to say that there are six 2-dimensional planes in 4-dimensional space. Rotation in a single 2-dimensional plane is associated with a single angle which is a real number like 30^0 or 43^0. Since we have six 2-dimensional planes in our space-time (three Euclidean planes and three space-time planes), and since we can rotate in every one of these 2-dimensional planes, rotation in our 4-dimensional space-time is associated with six real numbers which are the six angles (one for each 2-dimensional plane. We can wave our arms around in our 4-dimensional space-time because we can rotate our arms in all directions.

Looking at the quaternion rotation matrix, (6.9), above, we see only three real numbers, $\{b, c, d\}$, associated with the rotation matrix of this 4-dimensional quaternion space. In other words, in 4-dimensional quaternion space, we can rotate in only three of the six 2-dimensional planes – ultimately, this is because the finite group $C_2 \times C_2$ has only three C_2 sub-groups. The important bit is that we cannot wave our arms around in quaternion space. Quaternion space is not a geometric space like our 4-dimensional space-time.

Students of quantum physics will know that electrons can have only spin up or spin down and never anything in the middle. Electrons cannot arbitrarily rotate the axes of their spin planes.

Strange Rotations

Quaternion space exactly matches electron spin states in that electron spin states can rotate in only three 2-dimensional planes in a 4-dimensional space-time.

There is more. Because two 2-dimensional rotations combined together are another 2-dimensional rotation in a different rotational plane – try it with your hands – and because that different rotational plane has within it a little of the rotation plane perpendicular to the initial two 2-dimensional rotations, we need to be able to rotate in all 2-dimensional planes of a space to be able to rotate in two planes at the same time. We cannot rotate in all 2-dimensional planes within quaternion space, and so, in quaternion space, we cannot rotate in more than one 2-dimensional plane at a time. This means all rotations in quaternion space, although being of a 4-dimensional nature must also be of a 2-dimensional nature like (6.10). Thus, in quaternion space, it is not possible to rotate in any more than one 2-dimensional plane at a time, and so the 'rotational axis' cannot vary from being other than plus one or minus one. This is exactly what is observed for electron spin when we send electrons through a (Stern-Gerlach) machine to measure the directions of their intrinsic spin.

Summary of this chapter:
In this chapter we have met rotations which are strange to our eyes.

We first met spinor rotations:

a) Spinor rotations are the kind of rotations we find inside types of numbers (spinor spaces).

b) Spinor rotations are expressed as the rotation matrix which appears in the polar form of a type of numbers (division algebra[11]).
c) Spinor rotations are not rotations about an axis.
d) Spinor rotations are multi-angular rotations in that the trigonometric functions in the rotation matrix accept $(n-1)$ angles (real numbers) where n is the dimension of the space. The 2-dimensional spinor rotations do this, of course.
e) Spinor rotations are n-dimensional whereas we have experience of only 2-dimensional rotations.
f) Some spinor algebras, like the quaternions, have twice as much rotation as we think is morally proper. We call this double cover.

We secondly met rotation in space-time:

a) Because our 4-dimensional space-time is a single entity rather than a fibre bundle of two separate spaces, we can rotate in a 2-dimensional space-time plane, but this is misleading. The 2-dimensional space-time plane is a 2-dimensional spinor rotation sitting in a 4-dimensional space. We can rotate in 2-dimensional space-time because there exists a 2-dimensional type of numbers called the hyperbolic complex numbers which has a space axis and a time axis.
b) Rotation in space-time is change of velocity.

---

[11] Type of numbers or division algebra or spinor space – all the same thing.

c) Distance in 2-dimensional space-time is given by: $d^2 = t^2 - z^2$.

We thirdly met double cover rotation in a non-commutative spinor space, the quaternions:

a) Because the quaternions are a non-commutative spinor space, rotation through a given angle in a given direction will lead to two different positions depending upon whether we rotate on the left or rotate on the right.
b) Quaternion rotation is rotation in both directions at the same time rather than only clockwise or only anti-clockwise.
c) We can rotate in only three 2-dimensional planes in 4-dimensional quaternion space rather than the six 2-dimensional planes of our 4-dimensional space-time. This is why electron spin is quantitised.

It is definitely time for the pub.

Chapter 7

A Catalogue of Spinor Rotations

As the reader might have realised when they saw (2.7), things called finite groups are connected to the different types of numbers. There is a great deal of mathematics concerning finite groups, but we do not intend to study finite groups at any great depth in this book. However there are some things that every decent human being should know about finite groups.

Firstly, we give the standard mantra because the reader has already met or will doubtless meet this standard mantra some time. A finite group is a set of permutations. There are two ways in which we can order a black ball and a red ball. These ways are red black or black red. These two permutations are the finite group called C_2. The C stands for cyclic; in the case of the cyclic groups, the subscript gives the number of elements in the group; there are two elements in C_2. This is not the case with other types of groups. The number of elements in the group is called the order of the group.

Permutations are such that they can be multiplied together. Another example of the finite group C_2 is the two numbers $\{-1,+1\}$. We have:

$$-1 \times -1 = +1, \quad -1 \times +1 = -1 \\ +1 \times -1 = -1, \quad +1 \times +1 = +1 \qquad (7.1)$$

A Catalogue of Spinor Rotations

However we multiply the two elements of the group, we get an element of the group.

There are six ways in which we can order three different coloured balls. These six permutations are called the symmetric group S_3. Within S_3, there is a subgroup of the ways that three differently coloured balls can be arranged, RGB, GBR, BRG. We have introduced a green ball. This sub-group is called C_3.

Having given the standard mantra about permutations, we will begin again and proceed in a much clearer way.

Secondly, we choose an easier way to do finite group theory. An easier way is to simply write down all the square matrices of a particular size that have a single 1 in each row and a single 1 in each column. These are called permutation matrices because they are in one-to-one correspondence with permutations. We have:

$$\begin{bmatrix} 1 & 0 \\ 0 & 1 \end{bmatrix} \ \& \ \begin{bmatrix} 0 & 1 \\ 1 & 0 \end{bmatrix} \qquad (7.2)$$

This is the finite group C_2. We have:

$$\begin{bmatrix} 1 & 0 & 0 \\ 0 & 1 & 0 \\ 0 & 0 & 1 \end{bmatrix}, \begin{bmatrix} 0 & 1 & 0 \\ 0 & 0 & 1 \\ 1 & 0 & 0 \end{bmatrix}, \begin{bmatrix} 0 & 0 & 1 \\ 1 & 0 & 0 \\ 0 & 1 & 0 \end{bmatrix}$$
$$\begin{bmatrix} 1 & 0 & 0 \\ 0 & 0 & 1 \\ 0 & 1 & 0 \end{bmatrix}, \begin{bmatrix} 0 & 1 & 0 \\ 1 & 0 & 0 \\ 0 & 0 & 1 \end{bmatrix}, \begin{bmatrix} 0 & 0 & 1 \\ 0 & 1 & 0 \\ 1 & 0 & 0 \end{bmatrix} \qquad (7.3)$$

This is the finite group S_3. S_3 is an order six group. The top row of (7.3) alone is the finite group C_3. We could have written the S_3 group using 6×6 permutation matrices, and normally we would do this.

It is easy to find a type of numbers, a spinor space, of any dimension you choose. We choose 3-dimensions. We start with the permutation matrix that has all the 1's on the leading diagonal:

$$\begin{bmatrix} 1 & 0 & 0 \\ 0 & 1 & 0 \\ 0 & 0 & 1 \end{bmatrix} \quad (7.4)$$

We then choose permutation matrices of the same size that will fit into this, (7.4), in such a way that there is a 1 in every position:

$$\begin{bmatrix} 0 & 1 & 0 \\ 0 & 0 & 1 \\ 1 & 0 & 0 \end{bmatrix}, \quad \begin{bmatrix} 0 & 0 & 1 \\ 1 & 0 & 0 \\ 0 & 1 & 0 \end{bmatrix} \quad (7.5)$$

We multiply each of these matrices by a different real variable and add them. Then we use the exponential function to get the polar form:

$$\exp\left(\begin{bmatrix} a & 0 & 0 \\ 0 & a & 0 \\ 0 & 0 & a \end{bmatrix} + \begin{bmatrix} 0 & b & 0 \\ 0 & 0 & b \\ b & 0 & 0 \end{bmatrix} + \begin{bmatrix} 0 & 0 & c \\ c & 0 & 0 \\ 0 & c & 0 \end{bmatrix} \right) \quad (7.6)$$

A Catalogue of Spinor Rotations

$$= \exp\left(\begin{bmatrix} a & b & c \\ c & a & b \\ b & c & a \end{bmatrix}\right) \qquad (7.7)$$

$$= \begin{bmatrix} e^a & 0 & 0 \\ 0 & e^a & 0 \\ 0 & 0 & e^a \end{bmatrix} \begin{bmatrix} v_A(b,c) & v_B(b,c) & v_C(b,c) \\ v_C(b,c) & v_A(b,c) & v_B(b,c) \\ v_B(b,c) & v_C(b,c) & v_A(b,c) \end{bmatrix} \quad (7.8)$$

Note that $e^a = r$ is just a real number which is the (radial) distance of the number from the origin.

This is a 3-dimensional type of number. The 3-dimensional space associated with it is something no human has ever seen. For a start, there are no 2-dimensional sub-spaces in this space and so there are no 2-dimensional rotations. This is 3-dimensional rotation. The space also has a rather weird type of 3-dimensional reflection.[12] Actually, everything about this spinor space is weird to our eyes.

The functions in the rotation matrix, the nu-functions $\{v_A, v_B, v_C\}$, are the trigonometric functions of this space. The nu-functions, as are all trigonometric functions, are each a projection from the unit 'sphere' in this space on to an axis of this space.

In the case of the finite group C_2, the above procedure gives:

---

[12] See: Dennis Morris : Complex Numbers The Higher Dimensional Forms : ISBN: 9781508677499

$$\exp\left(\begin{bmatrix} a & 0 \\ 0 & a \end{bmatrix} + \begin{bmatrix} 0 & b \\ b & 0 \end{bmatrix}\right) = \begin{bmatrix} r & 0 \\ 0 & r \end{bmatrix}\begin{bmatrix} \cosh b & \sinh b \\ \sinh b & \cosh b \end{bmatrix}$$

(7.9)

This is 2-dimensional space-time. The rotation matrix is called the Lorentz transformation, and we effectively have the special theory of relativity from no more than a few 1's scattered about in a couple of small matrices – more later.

Scattering minus signs about within the matrix will give all the other types of numbers (division algebras) of the chosen dimension. (There is a precise mathematical way to calculate where we put the minus signs based on the need for multiplicative closure.) Of course, we have seen the other 2-dimensional type of numbers above.

So you see, every finite group, that is every set of square matrices with 1's scattered about them as described above, gives types of n-dimensional numbers which hold n-dimensional spinor rotations. I told you the maths was simple.

The order one group, C_1, is just the permutation matrix [1]. This gives the real numbers.

How many finite groups are there?
As within the real numbers there are prime numbers from which all other whole real numbers can be constructed, so there are 'prime' finite groups. These 'prime' finite groups are called simple finite groups. Possibly the greatest achievement of 20th century mathematics was the cataloguing of all the simple

finite groups. There are several families each containing an infinite number of simple finite groups, the cyclic groups are one example, and there are twenty six odd-balls called sporadic groups[13]. The number of families depends on how you count them, but some people count them to be three. Obviously, just as there are more whole real numbers than there are prime real numbers, there are more finite groups than there are simple finite groups.

Every one of the finite groups contains types of spinor rotations.[14]

The physical universe:
It seems that the whole of the physical universe, except possibly dark matter, is constructed from the finite groups C_1, C_2, $C_2 \times C_2$ and $C_2 \times C_2 \times C_2$. The finite group $C_2 \times C_2 \times C_2 \times C_2$ might have some input at extremely high energies well beyond the Large Hadron Collider, but we think not. These finite groups contain the types of numbers listed in (2.7). Of those, our concern is with only:

---

[13] The largest sporadic group is called 'The Monster'. It is connected to 32-dimensional string theory. Richard Borcherds won a Fields Medal for proving the connection to string theory. The connection is called the Monster Moonshine Theorem.

[14] Sci-fi writers take note. There really is such a thing as monster space.

C_1:   1 type of 1-dimensional space

C_2:   2 types of 2-dimensional space

$C_2 \times C_2$:   8 types of non-commutative 4-dimensional space (2 quaternion & 6 A_3 spaces)

$$(7.10)$$

We provide a list of these types of numbers.

The 1-dimensional real numbers:

$$[a] \qquad (7.11)$$

The 2-dimensional numbers:

$$\begin{bmatrix} r & 0 \\ 0 & r \end{bmatrix} \begin{bmatrix} \cosh b & \sinh b \\ \sinh b & \cosh b \end{bmatrix} \quad \& \quad \begin{bmatrix} r & 0 \\ 0 & r \end{bmatrix} \begin{bmatrix} \cos b & \sin b \\ -\sin b & \cos b \end{bmatrix}$$

$$(7.12)$$

The 4-dimensional numbers:

The two quaternion division algebras are:

$$\mathbb{H} = \begin{bmatrix} a & b & c & d \\ -b & a & -d & c \\ -c & d & a & -b \\ -d & -c & b & a \end{bmatrix} \qquad (7.13)$$

$$\mathbb{H}_{Anti} = \begin{bmatrix} a & b & c & d \\ -b & a & d & -c \\ -c & -d & a & b \\ -d & c & -b & a \end{bmatrix} \qquad (7.14)$$

A Catalogue of Spinor Rotations

The six A_3 division algebras are:

$$SSA^* = \exp\left(\begin{bmatrix} a & b & c & d \\ b & a & -d & -c \\ c & d & a & b \\ -d & -c & b & a \end{bmatrix}\right) \qquad (7.15)$$

$$SSA^*_{Anti} = \exp\left(\begin{bmatrix} a & b & c & d \\ b & a & d & c \\ c & -d & a & -b \\ -d & c & -b & a \end{bmatrix}\right) \qquad (7.16)$$

$$SAS = \exp\left(\begin{bmatrix} a & b & c & d \\ b & a & d & c \\ -c & d & a & -b \\ d & -c & -b & a \end{bmatrix}\right) \qquad (7.17)$$

$$SAS_{Anti} = \exp\left(\begin{bmatrix} a & b & c & d \\ b & a & -d & -c \\ -c & -d & a & b \\ d & c & b & a \end{bmatrix}\right) \qquad (7.18)$$

$$ASS = \exp\left(\begin{bmatrix} a & b & c & d \\ -b & a & -d & c \\ c & -d & a & -b \\ d & c & b & a \end{bmatrix}\right) \qquad (7.19)$$

$$ASS_{Anti} = \exp\left(\begin{bmatrix} a & b & c & d \\ -b & a & d & -c \\ c & d & a & b \\ d & -c & -b & a \end{bmatrix}\right) \quad (7.20)$$

A note:

The finite group $C_2 \times C_2$ can be thought of as the four pairs of numbers:

$$\begin{Bmatrix} +1 & -1 & +1 & -1 \\ +1 & -1 & -1 & +1 \end{Bmatrix} \quad (7.21)$$

We multiply these together as, for example:

$$\begin{pmatrix} +1 \\ -1 \end{pmatrix} \times \begin{pmatrix} -1 \\ -1 \end{pmatrix} = \begin{pmatrix} -1 \\ +1 \end{pmatrix} \quad (7.22)$$

The finite group $C_2 \times C_2 \times C_2$ can be thought of as the eight triples of numbers:

$$\begin{Bmatrix} +1 & +1 & +1 & -1 & +1 & -1 & -1 & -1 \\ +1 & +1 & -1 & +1 & -1 & +1 & -1 & -1 \\ +1 & -1 & +1 & +1 & -1 & -1 & +1 & -1 \end{Bmatrix} \quad (7.23)$$

These are multiplied together in a way analogous to the pairs of numbers shown in (7.22) above. The higher dimensional $C_2 \times C_2 \times ...$ finite groups can be represented in an analogous way.

A Catalogue of Spinor Rotations

Another taster:

We still have a long way to go, but we are well past the hardest part of this book. The reader deserves some reward. We will simply add the determinants (distance functions) of the six A_3 spaces. We have:

$$SUM \begin{cases} dist^2 = t^2 - x^2 - y^2 + z^2 \\ dist^2 = t^2 - x^2 - y^2 + z^2 \\ dist^2 = t^2 - x^2 + y^2 - z^2 \\ dist^2 = t^2 - x^2 + y^2 - z^2 \\ dist^2 = t^2 + x^2 - y^2 - z^2 \\ dist^2 = t^2 + x^2 - y^2 - z^2 \end{cases} = 2\left(3t^2 - x^2 - y^2 - z^2\right)$$

(7.24)

The 2 is just a scaling factor and the 3 is just the units in which we measure time or space. We can set them to 1; doing so gives:

$$d^2 = t^2 - x^2 - y^2 - z^2 \qquad (7.25)$$

This is called the A_3 expectation distance function. The reader will recognise this as the distance function of our 4-dimensional space-time.

From Where Comes the Universe?

Chapter 8

The Special Theory of Relativity

Kettles boil at $100^0 C$ regardless of whether the spout is pointing towards the north or towards the west. This blindingly obvious fact is part of the most important physical law in the universe. That law is that all physical processes are independent of the direction in which the apparatus is pointing. All the physics and the chemistry is the same regardless of direction, and all physical constants have the same values regardless of the direction in which they are measured. Light travels at the same velocity from west to east as it does from north to south.

We call this 'everything is the same in all directions' the isotropy of space.

Note: It is also the case that all physical processes are independent of their position in space. A kettle boils at $100^0 C$ a billion light years from Earth. We know this because we can observe the same physical processes happening in very distant galaxies as we observe in nearby galaxies.

It is also the case that all physical processes are independent of their position in time. Kettles boiled at $100^0 C$ a billion years ago[15] and will boil at $100^0 C$ a billion years into the future.

---

[15] Tea drinking is a long established tradition of our species.

The Special Theory of Relativity

In previous chapters, we have met different types of space. Do kettles boil at $100^0 C$ in every direction in every type of space? Yes! including in 2-dimensional space-time.

We met 2-dimensional space-time above. Recall that it fell out of the finite group C_2 and we called it the hyperbolic complex numbers. It is remarkable that, from no more than the two numbers $\{-1, +1\}$ and the real numbers, we get time and space.

Above, we saw that a direction in 2-dimensional space-time is a velocity, and so the isotropy of 2-dimensional space-time means that kettles boil at $100^0 C$ regardless of the velocity at which they are moving. A kettle moving at 10,000 miles per second boils at $100^0 C$ just as it does when it is stood still or moving at ten miles per second.

We have a name for this isotropy of 2-dimensional space-time. The name is the special theory of relativity.

Note: There are two theories of relativity. There is the special theory of relativity which is about velocity and with which we are concerned in this chapter and there is the general theory of relativity which is about gravity and with which we will be concerned in a later chapter.

The special theory of relativity says that every physical process works exactly the same at all velocities. Different velocities are just different directions in 2-dimensional space-time, and so the special theory of relativity simple says 2-dimensional space-time is isotropic (the same in all directions).

There are consequences of the isotropy of 2-dimensional space-time.

One of those consequences is that, while a stationary electric charge has a zero magnetic field, a moving electric charge has a non-zero magnetic field. This fact underlies the whole of electrical power generation and of electrical engineering and thus a huge part of our industry and civilisation. Let us take an electric charge with a zero magnetic field and rotate it in 2-dimensional space-time (that is change its velocity). Such an electric charge with a zero magnetic field is:

$$\begin{bmatrix} E & 0 \\ 0 & E \end{bmatrix} \quad (8.1)$$

We will rotate this electric charge through space-time angle χ - a change of velocity which corresponds to the real number χ. We rotate the electric charge simply by multiplying it by a rotation matrix:

$$\begin{bmatrix} \cosh \chi & \sinh \chi \\ \sinh \chi & \cosh \chi \end{bmatrix} \begin{bmatrix} E & 0 \\ 0 & E \end{bmatrix} = \begin{bmatrix} E\cosh \chi & E\sinh \chi \\ E\sinh \chi & E\cosh \chi \end{bmatrix}$$
(8.2)

The $E \sinh \chi$ is the magnetic field. And so, the whole of modern civilisation is possible because the numbers $\{-1, +1\}$ give rise to an isotropic 2-dimensional space-time.

There are other consequences of the isotropy of 2-dimensional space-time, but before we get to them, we will do a little hyperbolic trigonometry. Velocity is the ratio of distance to time:

$$v = \frac{\text{distance}}{\text{time}} \tag{8.3}$$

In 2-dimensional space-time, distance is the length along the space axis and time is the length along the time axis. The length along the space axis is the sinh( ) function and the length along the time axis is the cosh( ) function. Remember, trigonometric functions are just projections from the unit circle (a hyperbola in the case of 2-dimensional space-time) on to the Cartesian axes of the space. We therefore have, ignoring arbitrary units:

$$v = \frac{\sinh \chi}{\cosh \chi} \tag{8.4}$$

We will need to use the space-time Pythagoras theorem:

$$d^2 = t^2 - z^2 \tag{8.5}$$

This gives us:

$$\cosh^2 \chi - \sinh^2 \chi = 1 \tag{8.6}$$

Notice the minus sign.

Starting with (8.4) and squaring, we get:

$$\frac{\sinh^2 \chi}{\cosh^2 \chi} = v^2 \tag{8.7}$$

Using (8.6) gives:

$$1 - \frac{1}{\cosh^2 \chi} = v^2$$
$$\cosh \chi = \frac{1}{\sqrt{1-v^2}} \quad (8.8)$$

In fact, we need to adjust the units to suit our human choice of units. This gives:

$$\cosh \chi = \frac{1}{\sqrt{1-\frac{v^2}{c^2}}} \quad (8.9)$$

The reader might have seen this before. When Einstein first formulated the theory of special relativity, he used the form on the right of (8.9) rather than the hyperbolic trigonometric function on the left of (8.9). This form is still in common usage; your author prefers the simplicity of the hyperbolic trigonometric function form. We also get:

$$\sinh \chi = \frac{v}{\sqrt{1-\frac{v^2}{c^2}}} \quad (8.10)$$

Now let us consider a point in 2-dimensional space-time on the time axis and let us rotate this point:

$$\begin{bmatrix} \cosh \chi & \sinh \chi \\ \sinh \chi & \cosh \chi \end{bmatrix} \begin{bmatrix} t_0 & 0 \\ 0 & t_0 \end{bmatrix} = \begin{bmatrix} t_0 \cosh \chi & t_0 \sinh \chi \\ t_0 \sinh \chi & t_0 \cosh \chi \end{bmatrix}$$
$$(8.11)$$

$$= \begin{bmatrix} t_0 \dfrac{1}{\sqrt{1-\dfrac{v^2}{c^2}}} & \sim \\ \sim & t_0 \dfrac{1}{\sqrt{1-\dfrac{v^2}{c^2}}} \end{bmatrix} \qquad (8.12)$$

We see that time changes as we change velocity. If the velocity is $\dfrac{4}{5}c$, we have:

$$t_0 \dfrac{1}{\sqrt{1-\dfrac{v^2}{c^2}}} = t_0 \dfrac{1}{\sqrt{1-\dfrac{16}{25}}} = \dfrac{5}{4}t_0 \qquad (8.13)$$

This is called time dilation. To a stationary observer, the processes of the universe appear to unfold more slowly within moving objects. Clocks on board the Apollo space-craft which took astronauts to the moon lost three seconds compared to similar clocks on Earth. Length changes as well, but we will not look at that in this book; these things are adequately covered elsewhere[16].

We will look at time dilation from another point of view. Within 2-dimensional Euclidean space (the flat sheet of paper), the set of points a unit length from the origin is a Euclidean circle.

---

[16] See: Dennis Morris *Empty Space is Amazing Stuff*.

From Where Comes the Universe?

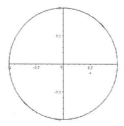

Within 2-dimensional space-time, the set of points a unit length from the origin is a hyperbolic circle.

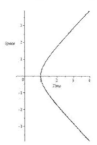

A hyperbolic circle is very different from a Euclidean circle. This is a hyperbola. The points on the line are a distance of one unit from the origin using the space-time distance formula (8.5). As we rotate in 2-dimensional space-time away from the horizontal axis, we move along the line that is the hyperbola. This is just the same as moving along the line that is the circle as we rotate in Euclidean space. The cosh( ) function is the distance along the horizontal time axis of a point on the line that is the hyperbola just as the cos( ) function is the distance along the horizontal axis of a point on the Euclidean circle.

The Special Theory of Relativity

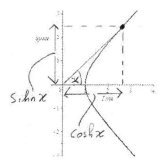

We see that, as we rotate in 2-dimensional space-time, the distance along the time axis increases – time stretches. This is time dilation.

There is much more to the theory of special relativity than is presented in this book, but we have not room for it here. The important point is that time-dilation and the other effects associated with special relativity are no more than geometry in the 2-dimensional space-time that we find inside the C_2 finite group.

The central point is that rotation in 2-dimensional space-time is a change of velocity. Of course, 2-dimensional space-time is a spinor space.

Before we finish this chapter, we will add one more note about the geometry of 2-dimensional space-time. Within Euclidean space (the flat sheet of paper) a straight line is the shortest distance between two points. Within 2-dimensional space-time, a straight line is the longest distance between two points. Of course, that's only fair. I did say the concepts were challenging.

Summary:
a) Change of velocity is just rotation in 2-dimensional space-time.
b) All physical processes work the same regardless of the direction of the apparatus in 2-dimensional space-time; this is regardless of the velocity of the apparatus.

An addendum – the expanding universe:

The reader might have heard that the universe is expanding and might think that distant galaxies are moving relative to ourselves. The reader might think that physical processes appear to us to be slowed in those distant galaxies, and so, in distant galaxies, stars burn more slowly and planets orbit more slowly and alien woman age more slowly and teenagers ... (perish the thought).

Nothing happens more slowly in a distant galaxy. This is because, although the galaxies are moving away from us, they are not really moving relative to us. The intervening empty space is stretching. How does empty space stretch? Look at the drawing of a hyperbolic circle (hyperbola) above. Time is on the horizontal axis. What happens to the width of the space as time increases? As time increases, the space gets wider. This is the expansion of the universe. The expansion of the universe is intrinsic to the nature of 2-dimensional space-time. This intrinsic property is passed on to the 4-dimensional space-time of our universe.

With a little co-ordinate geometry, it is possible to show that the expansion of the universe will appear to an observer to be accelerating as the universe gets older. This apparent acceleration of the rate of expansion has been observed.[17]

---

[17] See Dennis Morris: Upon General Relativity.

Chapter 9

Super-imposition

Within quantum physics, physicists have a concept which they call the expectation value of the result of an experiment. An expectation value is just an average value.

Within classical physics, the same experiment always produces the same result. Within quantum physics, the same experiment does not always produce the same result. A quantum physicist will predict the outcome of an experiment by considering thousands of identical experiments and predicting the average outcome. This average outcome is called the expectation value. Given enough experiments, expectation values are very accurate and very reliable, but they are nothing to do with the outcome of a single experiment.

Events in the sub-atomic world seem to us to be probabilistic and random. In one case, an electron might behave in a particular way but, in another identical case, the electron might behave differently. We find this lack of absolute determinism very strange, but perhaps it is even stranger that the probability of the electron behaving in a particular way is entirely determined. The probability of a particular outcome is always determined. Reluctantly, physicists have been driven to accept these observed facts. The probabilistic nature of the quantum universe has been observed trillions of times and the certainty of the expectation value concept has become part of our modern industrial world.

It is the act of making an observation (interacting with the observed object) which gives the expectation value. It is important to realise that the 'observer' need not be a human being but could be a physical object like an electron which interacts with the 'observed object'. Of course, the observation goes both ways, and we could think of the 'observed object' as the 'observer' and the electron as the object being observed.

Super-imposition of spinor spaces:

Super-imposition is no more than a generalisation of the expectation value concept. Instead of taking the average of a value, we are going to take average distance functions, average field equations, average tensors, and average spaces. The spaces of which we are going to take the average are the spinor spaces. The distance functions of which we are going to take the average are the spinor distance functions. We have done this already; see (7.24).

Again, it is the act of interaction (observation) which brings about the super-imposition.

We will be taking average algebras by super-imposing isomorphic division algebras. This means we will be adding together several copies of the same types of numbers. Our interest for the present is in only types of numbers (spinor spaces) of dimension four or less. In an earlier chapter, we have listed the spinor spaces in which we are interested, circa (7.10) . Notice that there are six A_3 spinor spaces.

Super-imposition - 2-dimensional:
There are two types of 2-dimensional numbers. There is only one copy of each type of 2-dimensional number that emerges from the finite group C_2.

Super-imposing one copy of a spinor space changes nothing, and so each of the two 2-dimensional spinor spaces come through super-imposition unscathed. In other words, the average of a single space is that single space.

Since each 2-dimensional space is unaffected by super-imposition, all the geometric properties such as 2-dimensional rotation, 2-dimensional angles, and 2-dimensional distance functions survive the super-imposition operation. There are other aspects of the 2-dimensional spaces which are more technical such as the 2-dimensional curl and the 2-dimensional inner product which also survive super-imposition.

The survival of both the 2-dimensional spaces means that we will see things like 2-dimensional rotations and 2-dimensional curls in our 4-dimensional classical universe. In fact, it will become apparent that a great deal of the geometric nature of our 4-dimensional universe is 2-dimensional geometry. Much of our 4-dimensional space-time is fabricated from the 2-dimensional spinor spaces; this would not be possible if the geometrical properties of the 2-dimensional spinor spaces had been destroyed by super-imposition.

Of course, the 1-dimensional real numbers survive super-imposition because there is only one copy of the 1-dimensional numbers which derives from the finite group C_1.

Super-imposition - 4-dimensional:
Our interest is in only the non-commutative 4-dimensional spinor spaces. There are two quaternion spaces and six A_3 spaces. The six A_3 spaces differ from each other in nothing more than that they are written in different bases.

We present an A_3 rotation matrix:

$$A_{3\,Rot} = \begin{bmatrix} \cosh(\lambda) & \frac{b}{\lambda}\sinh(\lambda) & \frac{c}{\lambda}\sinh(\lambda) & \frac{d}{\lambda}\sinh(\lambda) \\ \frac{b}{\lambda}\sinh(\lambda) & \cosh(\lambda) & -\frac{d}{\lambda}\sinh(\lambda) & \frac{c}{\lambda}\sinh(\lambda) \\ -\frac{c}{\lambda}\sinh(\lambda) & -\frac{d}{\lambda}\sinh(\lambda) & \cosh(\lambda) & -\frac{b}{\lambda}\sinh(\lambda) \\ \frac{d}{\lambda}\sinh(\lambda) & -\frac{c}{\lambda}\sinh(\lambda) & -\frac{b}{\lambda}\sinh(\lambda) & \cosh(\lambda) \end{bmatrix}$$

$$\lambda = \sqrt{b^2 - c^2 + d^2}$$

(9.1)

We see that the trigonometric functions of the A_3 spinor spaces are closely related to the trigonometric functions of the 2-dimensional space-time spinor space. This is a little deceptive; for mathematical reasons which we have not the time to discuss, of the three 2-dimensional rotation planes, two are similar to space-time rotations and one is similar to a Euclidean rotation.

> *Technical note:* The A_3 spinor spaces have one imaginary variable which is a square root of minus one and two imaginary variables which are square roots of plus one. As such, they are the division algebra forms of the Clifford algebras $Cl_{2,0} \cong Cl_{1,1}$.[18]

We will take the average of the A_3 spaces, the A_3 expectation space, by super-imposing the six A_3 spaces. (This is adding the six spaces, or adding the six division algebras if you prefer.) Oops! We cannot add mathematical objects written in different bases without destroying them. When we super-impose the A_3 spaces we destroy the algebraic structure of these spaces. One part of this algebraic structure is the multiplication operation. Another part of this algebraic structure is the 4-dimensional rotation. It is because the 4-dimensional rotation of the A_3 spaces is destroyed by super-imposition that we do not have 4-dimensional rotation in our classical universe. By super-imposition, we have broken the rotational symmetry of the A_3 spinor space.

So, with what are we left after we have super-imposed the A_3 spaces? We are left with four variables which fit together into 4-tuples of real numbers. We can think of them as points in a 4-dimensional space. The numbers have to be real because in the absence of a multiplication operation,

---

[18] See: Dennis Morris: The Naked Spinor.

Super-imposition

we cannot have imaginary numbers. The reader might think of this super-imposition as smashing the 4-dimensional structure of the A_3 spaces into four 1-dimensional spaces.

Mathematicians call a set of n-tuples of real numbers a manifold which they denote with the symbol $\mathbb{R}^n$. In the A_3 case, we have a 4-tuple of real numbers denoted by $\mathbb{R}^4$.

There is no sense of direction or parallel lines in a manifold; that was destroyed by super-imposition. There is no sense of rotation or angle in a manifold; that was destroyed by super-imposition. There is no sense of distance in a manifold; that was destroyed by super-imposition. Super-imposition of the A_3 spaces leaves us with nothing more than a set of 4-tuples of points – a manifold.

Well, not quite, there is something else that survives super-imposition. All spinor spaces are flat. There are different types of flat space; 2-dimensional space-time is flat, and a 2-dimensional sheet of paper is also flat.

At an infinitesimally small point, there is no difference between spaces that are written in different bases. At an infinitesimally small point, and this applies to all infinitesimally small points, we have a flat manifold. It is important to realise that this flatness does not extend beyond the infinitesimally small point because the A_3 spaces are each set in a different basis. However, the emergent manifold is flat at every infinitesimally small point. This is just like the surface of a sphere. At an infinitesimally small point, the surface of a sphere is exactly like a flat sheet of paper. Your author has drawn an artistically meritorious picture of this on the next page of this book.

Local flatness is what mathematicians simply call this local flatness at every infinitesimally small point of a space. They say that the surface of a sphere is locally flat. We see that super-imposition of the A_3 spaces leaves us with a locally flat manifold.

A locally flat manifold is a thing very different from a globally flat manifold. Locally flat manifolds can be globally curved. The surface of a sphere is globally curved.

The expectation distance function:
But wait, super-imposition of the distance functions of the A_3 spaces produces a distance function. A distance is just a number, and numbers can be added together without any problems. We have seen the A_3 expectation distance function above, (7.24).

Thus, we have a 4-dimensional locally flat manifold with a distance function.

A flat tangent space:
The theory of general relativity is based upon an area of mathematics called Riemann geometry. Riemann geometry deals with curved spaces such as the surface of a sphere. An essential part of Riemann geometry is the concept of a flat tangent space. This is a flat space which touches the curved space at only

an infinitesimally small point. We illustrate this with the surface of a sphere.

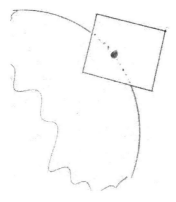

Because the surface of a sphere is 2-dimensional, the flat tangent space is 2-dimensional. In the case of a 4-dimensional manifold, the flat tangent space would have to be 4-dimensional.

We need the curved space to be locally flat to allow us to attach a flat tangent space to it, but we have seen that local flatness emerges from super-imposition, and so we can, if we can find one, attach a flat tangent space to the emergent manifold at each point. If we can do that, since we already have a Riemann distance function, we will have Riemann geometry; we still have to find an appropriate flat tangent space.

Well, the flat tangent space has the distance function which emerged from super-imposition which, in the A_3 case, is:

$$d^2 = t^2 - x^2 - y^2 - z^2 \tag{9.2}$$

Unfortunately, the only flat spaces are the spinor spaces, and there are no 4-dimensional spinor spaces with this distance function. (This is another way of saying that our 4-dimensional space-time is not a type of numbers – a spinor space.) That's it, we conclude that the classical universe can't exist. But then we realise that the distance function (9.2) can accommodate the 2-dimensional distance functions of the 2-dimensional spinor spaces. By this, we mean that if any two of the four variables in the 4-dimensional distance function (9.2) are zero, then we have one or other of the two 2-dimensional spinor distance functions. In fact, the 4-dimensional A_3 distance function (9.2) can accommodate six of the 2-dimensional spinor distance functions because there are six ways to choose a pair of variables from the four variables of the above 4-dimensional A_3 expectation distance function. Three of these ways are Euclidean:

$$d^2 = x^2 + y^2$$
$$d^2 = x^2 + z^2 \quad\quad (9.3)$$
$$d^2 = y^2 + z^2$$

and three of these ways are space-time:

$$d^2 = t^2 - x^2$$
$$d^2 = t^2 - y^2 \quad\quad (9.4)$$
$$d^2 = t^2 - z^2$$

This is, of course, exactly what we observe in our universe. We can rotate in three different planes and we can change velocity in three different spatial directions.

The universe can fabricate a flat tangent space by fabricating together six 2-dimensional spinor spaces. The expectation distance function determines just how this fabrication fits together. Along with the tangent space, the fabrication brings other technical aspects of the 2-dimensional spaces. A 4-dimensional inner product is similarly fabricated from six 2-dimensional inner products. The 4-dimensional inner product is important because it leads to a mathematical object called the metric tensor which is a very important part of Riemann geometry.

But there is no affine connection:
Well! we have made a lot of progress towards seeing how our 4-dimensional space-time emerges from the super-imposition of the A_3 spinor spaces, but we have not yet finished. The emergent space has distance function and 2-dimensional angles, but it does not have a sense of direction; that is, it does not have a definition of what it means to say that two lines are pointing in the same direction – that two lines are parallel to each other. We need a geometric object called an affine connection which determines direction within a space. We will derive the requisite affine connection in a later chapter of this book, but for now, let us rest upon our laurels and take a look at the super-imposition of the other spinor spaces.

What about the quaternions?

If we can super-impose the six A_3 spaces, we can super-impose the two quaternion spaces. That is correct. Why do we not see a quaternion space about us? Super-imposition of the two quaternion spaces smashes the quaternion spaces just as it smashed the A_3 spaces. The expectation distance function of the quaternion spaces is:

$$d^2 = a^2 + b^2 + c^2 + d^2 \qquad (9.5)$$

There are no minus signs in the quaternion expectation distance function. This means that the quaternion distance function cannot support rotations in space-time. It cannot support changes of velocity, and there is no concept of time. With no concept of time, we think it would not be observable to us, but we are not certain of this. We think the quaternions are connected to the weak nuclear force, but we are not sure of this either.

Interestingly, the absence of time within the quaternion emergent expectation space allows instantaneous communication at a distance. Such instantaneous communication has been observed within quantum physics and is called non-locality. Such non-locality is connected to the instantaneous collapse of a spatially extensive wave-function which the reader might encounter in books on quantum mechanics. Non-locality is also part of the quantum entanglement of two spatially separated photons of light. Quantum entanglement, and non-locality in general, are of great interest to physicists because they imply instantaneous communication over great distances. Such instantaneous communication over distance cannot happen in our 4-dimensional space-time, but it seemingly does happen in the quaternion emergent expectation space. One wonders if aliens from different galaxies communicate with each other through the quaternion emergent expectation space[19].

What about the 8-dimensional spinors?

The reader might think that what we have done with the six A_3 spaces from the $C_2 \times C_2$ finite group, we can do with, say, a set of 8-dimensional spinor spaces from the $C_2 \times C_2 \times C_2$ finite group. A very sensible observation. When we try to do this, nothing works properly. It is impossible to fabricate an emergent expectation space from any type of 8-dimensional spinor spaces that arise from the $C_2 \times C_2 \times C_2$ finite group. We will illustrate why this is so using the 4-dimensional case.

In 4-dimensional space, there are:

$$\frac{N(N-1)}{2} = \frac{4 \times 3}{2} = 6 \qquad (9.6)$$

2-dimensional planes.

There are three possible different 4-dimensional distance functions which support 2-dimensional rotation[20]:

---

[19] Perhaps there is an intergalactic internet up and running, and all we need to do is discover how to log on to it.
[20] A distance function will support 2-dimensional rotation if it reduces to a 2-dimensional distance function when only two of the variables are not zero.

$$d^2 = t^2 + x^2 + y^2 + z^2$$
$$d^2 = t^2 + x^2 + y^2 - z^2 \quad (9.7)$$
$$d^2 = t^2 + x^2 - y^2 - z^2$$

(Note that $d^2 = t^2 - x^2 - y^2 - z^2$ is equivalent to $d^2 = t^2 + x^2 + y^2 - z^2$.)

In any dimension, the distance function with all the same signs will have $\dfrac{N(N-1)}{2}$ Euclidean rotational planes. In any dimension, the distance function with only one different sign will have $(N-1)$ space-time rotational planes and thus:

$$\frac{N(N-1)}{2} - (N-1) = \frac{(N-1)(N-2)}{2} \quad (9.8)$$

Euclidean rotational planes. In any dimension, the distance function with two different signs will have $2(N-2)$ space-time rotational planes and thus:

$$\frac{N(N-1)}{2} - 2(N-2) = \frac{N^2 - 5N + 8}{2} \quad (9.9)$$

Euclidean rotational planes. Thus, a 4-dimensional emergent expectation space might be of the forms:

6 Euclidean rotations
3 Euclidean rotations & 3 Space-time rotations
2 Euclidean rotations & 4 Space-time rotations
$$(9.10)$$

Clearly, our space-time is the 3 Euclidean rotation and 3 space-time rotations one. If it were the case that the A_3 expectation distance function was of the form $d^2 = t^2 + x^2 - y^2 - z^2$, then we would have a space-time with 2 Euclidean rotations and 4 space-time rotations. The maths does not give this.

We see that there are no possible 4-dimensional spaces which can be fabricated from the 2-dimensional spaces with, say, five Euclidean rotations and one space-time rotation or, say, 1 Euclidean rotation and five space-time rotations.

If we examine the 8-dimensional spinor spaces from the $C_2 \times C_2 \times C_2$ finite group, we find that although individual pairs of variables can support a type of 2-dimensional spinor space, the relative numbers of the different types of 2-dimensional spaces are such that they cannot be fabricated into an 8-dimensional, or even a 4-dimensional, emergent expectation space. Since the 8-dimensional spinor spaces cannot support an 8-dimensional emergent expectation space, the 16-dimensional spinor spaces from the $C_2 \times C_2 \times C_2 \times C_2$ finite group cannot be fabricated into an emergent expectation space.

What about the 3-dimensional spinors?
There are two 3-dimensional spinor spaces which, like the 2-dimensional spinor spaces, survive super-imposition because only one copy of each of them derives from the C_3 finite group. Can emergent expectation spaces be fabricated from these 3-

dimensional spaces? It seems not, but research is ongoing in this area.

Conclusion:

The upshot of all this is that, and we still do not have a mathematical proof of this, the 4-dimensional emergent expectation space of the A_3 spaces and the emergent expectation space of the quaternion spaces seem to be the only possible expectation spaces to emerge from the spinor spaces. Our classical universe is the only possible classical universe, but it does have non-locality within it.

Summary:
   a) We super-imposed the six A_3 spinor spaces, and we destroyed the algebraic structure of those spaces including 4-dimensional rotation. We were left with a locally flat 4-dimensional manifold, $\mathbb{R}^4$.
   b) We super-imposed the six A_3 distance functions, and we got an emergent expectation distance function which matches the observed distance function of our 4-dimensional space-time.
   c) The 2-dimensional spinor spaces survive super-imposition because only one copy of each derives from the C_2 finite group.
   d) We fabricated a flat tangent space with the emergent expectation distance function from six copies (three of each type) of the 2-dimensional spinor spaces.

Super-imposition

e) The fabrication of the flat tangent space brings with it the fabrication of a 4-dimensional inner product which will produce the metric tensor.
f) We still need an affine connection, a sense of what it means for two directions to be parallel, to form our 4-dimensional space-time.
g) We found non-locality in the quaternion emergent expectation space.

An important point:
The 2-dimensional space-time of special relativity is a spinor space. The 4-dimensional space-time in which general relativity is formulated is a fabricated emergent expectation space. These are two very different types of space.

Chapter 10

Fibre Bundles and QFT

QFT stands for quantum field theory. What is a field?

We have all heard of gravitational fields or magnetic fields. In one sense, a field is something we assume exists because we do not like the idea of non-mechanical action at a distance. The Earth pulls on the moon without touching it; how does it do this? Either we allow that objects can act upon each other at a distance without touching or we invent a field which connects the two objects in some mysterious way.

An alternative view of a field is that it is a kind of space that is different from the underlying 4-dimensional space-time of our universe. The field space is spatially extensive –it spreads out over our 4-dimensional space-time – but it is not the same as our 4-dimensional space-time. Being a different space, it will have a different distance function, different angles, and a different type of rotation. Forces will arise as antagonism between the different distance functions of the two spaces or as antagonisms between the different types of rotations of the two spaces.

Another view is that a field is a fibre bundle with 4-dimensional space-time as an underlying space and a different type of space fixed to that underlying space at every point of the underlying space. This is the QFT view. This is a slightly different view from

Fibre Bundles and QFT

the one presented immediately above because it uses an infinite number of infinitesimally small copies of the field space rather than only one spatially extensive copy of the field space. However, it becomes the same as the view presented immediately above if we allow the origin of the field space to float freely over the underlying space rather than be fixed at a particular point of the underlying space.

Clearly, we can have more than one type of field over a single underlying 4-dimensional space-time if we use more than one type of field space.

Within QFT, the field spaces are quantum fields and are spinor spaces over the underlying 4-dimensional space-time of our universe as one big fibre bundle with many field spaces.

Conventional QFT does not need the complete spinor space; conventional QFT extracts from the spinor space only the unit sphere of rotation. These spheres of unit distance from the origin are called Lie groups[21].

We will look at QFT in a slightly unconventional way in that we will consider the whole spinor spaces rather than extracting bits of them

Fibre bundles again:
The reader will recall that a fibre bundle is one or more types of space fixed to every point of an underlying space. We gave the example of

---

[21] There is a technical complication which means the spheres of unit distance are like Lie groups rather than the actual Lie group, but this need not trouble us.

Newtonian space in which the underlying space is 1-dimensional time and the attached space is 3-dimensional Euclidean space.

We choose our 4-dimensional space-time to be the underlying space. We are going to fix spinor spaces to our 4-dimensional space-time. There are a limited number of spinor spaces in which we are interested.

The photon field:
We begin by fixing a copy the 2-dimensional Euclidean complex numbers (the flat sheet of paper) at every point in our space-time. The reader might envisage this as getting an infinitude of flat sheets of paper, drawing a pair of Cartesian axes on each sheet of paper, and using a drawing pin to pin the flat sheets of paper through the origin to every one of the infinite number of points in our 4-dimensional space-time. We are fixing a 2-dimensional space to our 4-dimensional space-time.

How do we orientate the axes of the sheets of paper? Do we arrange them such that the horizontal axes are all horizontal and the vertical axis are all vertical and match each other? Well, we humans probably would so align the axes, but nature does not. Because there is no mathematical connection between the fixed space and the underlying space, there is nothing to determine how the axes of the fixed space should be oriented. Because there is no mathematical connection between different copies of the fixed spaces, there is no communication between them that would lead to their axes being aligned uniformly. In short, in nature, the axes of the fixed spaces are oriented at random.

Fibre Bundles and QFT

We cannot cover the mathematical details within this book, but within QFT the fixing of copies of the 2-dimensional complex plane (sheets of paper) at every point of our 4-dimensional space-time gives rise to the electro-magnetic field (the photon field). The strength of the electro-magnetic field between two points in our 4-dimensional space-time depends upon the alignment the axes of the complex planes at the two points. In the alignment of the axes is the same, there is no electro-magnetic field. If the alignment of the axes is different, there is an electro-magnetic field between the two points whose strength depends upon the mismatch in the alignment of the axes. The technical phrase is local variation of the phase of the wave function. According to QFT, this is the origin of forces within the quantum universe. There is a nomenclature issue in that quantum physicists refer to rotation in the Euclidean complex plane as the Lie group $U(1)$.

The above single paragraph does not seem to do justice to the achievement of QFT. None-the-less, it does express the essence that the electro-magnetic force arises from forming a fibre bundle by fixing copies of the 2-dimensional Euclidean spinor space to our 4-dimensional space-time.

The shot-put force:
We now will do with the hyperbolic complex numbers (2-dimensional space-time) exactly what we have just done above with the Euclidean complex numbers[22]. We form a fibre bundle in which, again,

---

[22] This is not an included part of conventional QFT, but it ought to be.

our 4-dimensional space-time is the underlying space and copies of the hyperbolic complex numbers are fixed through the origin at every point to that underlying space. Again, there is nothing to correlate the alignment of the axes of the fixed spaces, and so the alignment of the axes can vary from point to point in our 4-dimensional space-time. But we know that different angles in the hyperbolic complex numbers correspond to different velocities. In other words, because the alignment of the axes of the hyperbolic complex number spaces varies from point to point in our 4-dimensional space-time, the velocity varies from point to point in our 4-dimensional space-time.

As a particle moves from one point in our 4-dimensional space-time to another point in our 4-dimensional space-time, its velocity will change if the axes of the fixed hyperbolic spaces at those two points are not aligned. Such a change of velocity is associated with an acceleration. An acceleration is associated with a force. We have a force associated with a change of phase of the fixed hyperbolic complex number spaces. The proportionality constant between the force and the acceleration is the inertial mass of the particle.

Oh! we have a force associated with the non-alignment of axes of the fixed spinor space in the fibre bundle. We also have inertial mass associated with a 2-dimensional space-time spinor rotation.

The quaternions and the weak nuclear force:
For the non-layman, we point out the technical fact that the commutation relations of the quaternions

Fibre Bundles and QFT

are isomorphic to the Lie group $SU(2)$. This technicality need not concern the layman.

We will do with the 4-dimensional quaternion space exactly as we have done with both the 2-dimensional spinor spaces. We will form a fibre bundle with our 4-dimensional space-time as the underlying space and an infinite number of copies of the quaternion space as the fixed space. Again, the orientation of the axes of the quaternion space will vary from point to point within our 4-dimensional space-time. This local variation of quaternion phase is known in QFT as local variation of $SU(2)$ phase. Within QFT, this local variation of quaternion phase gives rise to the weak nuclear force.

It seems that the local variation of anti-quaternions phase gives rise to the anti-weak nuclear force which we presume is indistinguishable from the weak nuclear force at quantum level.[23]

The strong nuclear force:
Within conventional QFT, the strong nuclear force arises as a consequence of local variation of the phase of the $SU(3)$ Lie group over the underlying 4-dimensional space-time. $SU(3)$ is an 8-dimensional Lie group. The lay person does not need to know the details of $SU(3)$ or what commutation relations are,

---

[23] Our presumption might be incorrect. This might be associated with the violation of parity by the weak force, but we do not yet understand this properly.

and so your author asks for the patience of the reader over the next paragraph which is a little technical.

At this point, we fail to agree with standard QFT. We have the 8-dimensional $C_2 \times C_2 \times C_2$ spinor spaces which we can use to form fibre bundles. There are three non-isomorphic types of non-commutative 8-dimensional $C_2 \times C_2 \times C_2$ spinor spaces which we could use, but none of these three have the commutation relations of $SU(3)$. Whereas, in four dimensions, we had a match between the $SU(2)$ commutation relations and the quaternion commutation relations, in eight dimensions, there is no match between the commutation relations of a spinor space and the $SU(3)$ commutation relations. Your author thanks the reader for their patience.

Each space of the many spaces of one type of these three non-commutative 8-dimensional spinor spaces 'folds' into either a 4-dimensional quaternion or a 4-dimensional anti-quaternion. The 'folding' is that the eight variables pair together into four pairs of variables. This is not like the 'rolled up' missing dimensions of string theory where the missing dimensions are just assumed to be rolled up. This is proper mathematical folding. Such folding is not properly understood, but we opine that this space will appear in 4-dimensional space-time to be no more than a quaternion or anti-quaternion.

Each space of the many spaces of another type of these three non-commutative 8-dimensional spinor spaces 'folds' into one or the other of the six 4-dimensional A_3 spaces.

This leaves one type of non-commutative 8-dimensional space which we think might give rise to the three colour forces of the strong nuclear force. Because these 8-dimensional spinor spaces do not have the same commutation relations as $SU(3)$, instead of predicting eight gluons, we predict six gluons. We think this is nice because there are exactly six things for gluons to do.

This is an area of active research, and we do not yet understand the strong nuclear force properly.

The A_3 spinors spaces:
The reader might have realised that we have not considered forming a fibre bundle with the A_3 spaces.

Until now, your author has omitted to point out something of great importance. In order to be able to say that the axes of a spinor space are oriented differently at different points in the underlying space of a fibre bundle, it is necessary to have a sense of direction within that underlying space – a sense of what parallel means. We have such a sense of direction within our 4-dimensional space-time, but, so far, we have seen no such sense of direction in the manifold which emerges from the super-imposition of the A_3 spinor spaces. Without a sense of what parallel means, technically called an affine connection, within the A_3 emergent expectation space, we are stuffed. We need an affine connection within the A_3 emergent expectation space, for QFT to work over this emergent expectation space.

What about the 3-dimensional spinors?
Why do we not use one of the four commutative 3-dimesional spinors spaces as a fixed space to form a fibre bundle over our 4-dimensional space-time? Come to think of it, what about the commutative 4-dimensional spinor spaces that derive from the finite group C_4 or, indeed, the eight commutative spinor spaces that derive from the finite group $C_2 \times C_2$ alongside the eight non-commutative spinor spaces?

Good question, the reader is obviously a particularly alert individual.

We allow the orientation of the fixed spaces to vary from point to point in a fibre bundle, but what do we mean by orientation? Do we measure orientation using a 2-dimensional Euclidean angle such as we see on a flat sheet of paper, or do we measure it with, say, a 3-dimensional angle as we find in the 3-dimensional spinor spaces or with a 4-dimensional angle? Our 4-dimensional space-time is a fabrication of 2-dimensional spinor spaces; it contains only 2-dimensional rotations (both types), and so any variation of orientation must be of a 2-dimensional nature. Variation of orientation in a 3-dimensional space is not of a 2-dimensional nature. If we used a 3-dimensional spinor space as a fixed space over our 4-dimensional space-time to form a fibre bundle, the variation of 3-dimensional orientation would be meaningless. Similarly the same is true of the other plethora of spinor spaces.

If we are to measure variation of orientation with 2-dimensional angles (both types), then the fixed spinor spaces must be such that they support 2-dimenstional rotations. This means that they must have pairings of variables which are tied together by one of the two 2-dimensional distance functions:

$$d^2 = t^2 - z^2$$
$$d^2 = x^2 + y^2$$
(10.1)

Only the 2-dimensional spinors, the quaternions, the A_3 spinor spaces, and the 8-dimensional $C_2 \times C_2 \times C_2$ spinor spaces can do this[24]. That is why we use only these spinor spaces in forming our fibre bundle.

Summary:
    a) We have seen that, other than gravity, the fundamental forces of physics arise from spinor spaces of locally varying orientation being fixed to the underlying 4-dimensional space-time as one big fibre bundle.
    b) We have used every possible spinor space there is which is commensurate with the 2-dimensional nature of the fabrication of our observed 4-dimensional space-time except the A_3 spinor spaces.
    c) We need an affine connection (a sense of what parallel means) within the underlying space of a fibre bundle to be able to measure local variation of orientation.

---

[24] The higher dimensional $C_2 \times C_2 \times ...$ can also do this, and so there might be more physics at extremely high energies.

Chapter 11

General Relativity

The theory of general relativity, usually abbreviated to GR, was published by Albert Einstein in 1915. It is a theory of the gravitational force which replaced the gravitational theory of Isaac Newton.

As judged by experimental observation, GR is a perfectly successful theory; it has perfectly matched every experimental test to very great accuracy.

The first postulate of GR:
GR is a theory based on three postulates. The first postulate of GR is the equivalence principle which states that:

> *Equivalence Principle:* An observer in free-fall in a uniform gravitational field is in an inertial reference frame. (An observer stood upon the surface of the Earth is in an accelerated reference frame.)

There is an often unstated assumption within both special relativity and GR that inertial reference frames are the fundamental reference frames of the universe. Special relativity is based on the idea that observers moving at constant velocities relative to each other are each in an inertial reference frame. GR extends this idea to include observers in free-fall in a uniform gravitational field; such observers are also in inertial reference frames.

A consequence of observers in free-fall being in an inertial reference frame is that observers who are within a gravitational field but are not in free-fall are not in an inertial reference frame. Thus, we have the view that a gravitational field is equivalent to an accelerated reference frame. Hence the 'equivalence' in equivalence principle.

Clearly, the reference frame arbitrarily chosen by a mortal observer ought not to affect the physics of the universe. Using this idea, from the equivalence principle, we can deduce the gravitational bending of light rays, gravitational red-shift and gravitational time dilation. If we take the view that light moves in a straight line, as it does in inertial reference frames, then, from the gravitational bending of light rays, we can deduce that space-time is curved.

We can view the equivalence principle part of the theory of GR as being simply a statement of the consequences of a self-evident truth. If the self-evident truth is true, we can be certain that the consequences of it are true. Of course, self-evident truths are sometimes not true.

The second postulate of GR:
The second postulate of GR is the field equations which describe how space-time curves in response to mass-energy. GR associates gravity with the curvature of space-time. In the modern view, GR does not postulate that our 4-dimensional space-time is embedded in a curved way within a higher dimensional flat space. In the modern view, GR takes the curvature of space-time to be intrinsic curvature such as, for example, the curvature of a spherical

surface (think surface of the Earth) which cannot be flattened. An intrinsically curved space contains its own intrinsic geometry independently of any higher dimensional space in which the space sits just as the surface of a sphere contains its own intrinsic geometry independently of any higher dimensional space in which the sphere sits.

The GR field equations cannot be deduced from the equivalence principle; they are a separate postulate of the theory of GR. Furthermore, the Riemann mathematical structure of the curved space-time within which the field equations are expressed is assumed by the theory rather than deduced.

We can view this second part of GR as a model of gravity constructed by humankind from thin air; how good the model is we determine by comparison with reality. We can never be certain that a model is true; a model is no more than very useful.

The third postulate of GR:
The third postulate of GR is the existence of the energy-momentum tensor. The energy-momentum tensor is a 4×4 matrix. This is to say that the 'cause' of gravity is not a scalar field (a single real number) of simply mass but is a tensor field whose components are energy and momentum. This reflects the equivalence of mass and energy as expressed by the famous equation $E = mc^2$ of special relativity. Of course, Isaac Newton saw gravity as being 'caused' by only the mass of a body.

Within GR, we must accept that energy gravitates. An example of this is the pressure energy within a collapsing supernova. As the supernova shrinks in size the internal pressure increases and so the energy stored in that pressure increases and so the gravitational force associated with that energy increases and so the star is squeezed further and the pressure increases even more. The mathematics is such that eventually the gravitational force 'caused' by the pressure is so great that it overcomes the pressure force and so a black hole is formed.

What GR does not explain:

Although GR is perfectly successful as a theory that allows us to predict and understand phenomena, the structure of general relativity is not perfect in at least five ways:

a) The field equations of GR have to be guessed[25]; they cannot be deduced within the theory.
b) GR must assume that space-time is 4-dimensional; this cannot be deduced from the theory.
c) GR must assume the distance function of our space-time; this cannot be deduced from the theory.
d) GR assumes that space-time is of a Riemann nature including local flatness and the presence of two types of 2-dimensional angles and the associated rotations and inner products.

---

[25] Einstein made three guesses. The first two guesses did not fit with observation; the third guess did fit with observation, and so we have his third guess for the field equations of GR.

e) GR assumes the existence of the energy-momentum tensor.

Further, although GR encompasses and includes the special theory of relativity, GR stands separate from quantum field theory, QFT. The mathematical structure of GR does not allow gravity to be quantitised in the same way that classical electromagnetism is quantitised, and there is no sign of Planck's constant in GR. For anything other than microscopic physics, this is not a problem, but it is expected that we will need a quantum theory of gravity to deal with very intense microscopic gravitational phenomena.

Our inability to deduce the field equations of GR means that field equations other than the GR field equations might be the correct field equations rather than the ones we have guessed. Of course, any set of field equations must accord with observation, but that does not eliminate all possible proposed field equations. Known to your author, there are two other proposed sets of field equations which accord with observation. One of these is the Brans-Dicke field equations of the scalar tensor theory of gravity and the other of these is the field equations of Elie Cartan's torsoin theory of gravity. We prefer the Einstein field equations for no reason other than they are simpler than the other two proposals, but we cannot eliminate the other two sets of proposed field equations by observation.

GR gravity compared to the other forces:
The GR theory of the gravitational force differs from the other forces of nature in that it is a force within

space-time rather than over space-time. Given the observation that the gravitational force provides the same acceleration to different masses at the same location, we might have suspected that gravity could not be a field over space-time. Attempts have been made to write gravitation as a scalar field, a vector field, or a tensor field over space-time, but these attempts fail to match observation.[26] We are thus quite certain that the gravitational force is not over space-time. However GR does not explain why gravity is within space-time whereas the other forces of physics are over space-time.

The structure of GR space-time:
Riemann geometry is the mathematics that underpins the GR field equations and the concept of intrinsic curvature. Riemann geometry was developed by Berhard Riemann (1826-1866) in 1854; it is not derivable from the equivalence principle.

The Riemann nature of the space-time of GR is a construction of several parts and several assumptions.

a) A manifold which is a set of n-tuples of real numbers.
b) Local flatness of that manifold.
c) A type of flat space to be used as a tangent space to the manifold. There are different types of flat space.
d) An affine connection; this is a definition of the parallel transport of a vector.

---

[26] See: John A Peacock - Cosmological Physics pg. 27

e) A metric tensor; this is a measure of distance and angle. This is just a set of inner products of basis vectors.
f) Two types of 2-dimensional angles and the associated inner products. These are the Lorentz boost angles in 2-dimensional space-time and the Euclidean angles of a 2-dimensional flat plane.

The metric tensor is constructed from the fabrication of 2-dimensional inner products. The Riemann curvature tensor is constructed from the metric tensor by differentiation, but we cannot differentiate without an affine connection. The Ricci tensor and the Ricci scalar are formed by contracting the Riemann tensor. The Einstein tensor is constructed by subtracting half the product of the metric tensor with the Ricci scalar from the Ricci tensor.

We see that the Einstein tensor is derived from the metric tensor provided we have an affine connection so that we can differentiate.

In short, all the important tensors of GR except the energy-momentum tensor are derived from the two types of 2-dimensional spinor angles and the associated 2-dimensional inner products and the two 2-dimensional spinor distance functions provided we have an affine connection. Of course, an affine connection is a definition of parallel lines.

GR also assumes the existence of the energy-momentum tensor. Putting, by guesswork[27], the

---

[27] The guesswork is educated. These are the only two tensors we have that have zero divergence – which is how Einstein guessed their equality.

Einstein tensor equal to the energy-momentum tensor produces the field equations of GR.

The existence of locally flat manifolds, flat tangent spaces, affine connections, metric tensors, and 2-dimensional angles are assumptions of Riemann geometry.

Summary:

Riemann space-time is based on assuming:

- a) A manifold
- b) Local flatness of the manifold
- c) Flat tangent spaces
- d) An affine connection
- e) A metric tensor based on inner products within the two 2-dimensional spinor spaces.
- f) 2-dimensional angles

GR is based on:

- a) The equivalence principle
- b) The GR field equations written in Riemann space-time
- c) The energy-momentum tensor

Chapter 12

The Affine Connection

The manifold which emerges from the super-imposition of spinor spaces has within it no sense of direction. Within the manifold, without any sense of direction, it is impossible to say that two lines are parallel.

Within a flat space like a sheet of paper, parallel lines never meet. They are what you were told they are in school. Within a curved space, like the surface of a sphere, parallel lines either meet, as on a sphere, or grow further apart, as on a hyperbolic surface. We can measure the amount of curvature in a space by measuring how similar parallel lines in that space are to parallel lines in a flat sheet of paper (or parallel lines in flat 2-dimensional space-time). The intrinsic curvature of a space is deviation from the flat nature of parallel lines.

The reader will recall that we previously formed a fibre bundle from our underlying 4-dimensional space-time and several spinor spaces but that we did not include the A_3 spinor spaces in this fibre bundle. We now consider the A_3 spaces.

We are going to form a fibre bundle with the A_3 emergent manifold and the six A_3 spinor spaces. Now, in each A_3 spinor space there is a set of axes. Within our fibre bundle, we have, at each point of the

The Affine Connection

underlying space six randomly oriented A_3 spinor spaces. The six orientations add by super-imposition to form a single orientation. We now have an underlying manifold and a single A_3 orientation at each point of the manifold.

We allow that the single super-imposed A_3 orientation varies from one point to another within the underlying manifold. But wait! the reader cries. It is meaningless to say that the orientation varies from point to point because there is no sense of direction within the underlying manifold. Of course, the reader is correct.

Now comes the big leap. The locally varying A_3 orientation induces a sense of direction into the underlying manifold. Basically, the different A_3 orientations are all parallel.

You might think that these different orientations point in different directions and so they are not parallel, but, when you so think, you are assuming a pre-determined sense of direction within the underlying manifold. There is no such pre-determined sense of direction within the underlying manifold against which you can compare the locally varying A_3 orientations to say that they point in different directions. The only meaningful conclusion is that the locally varying A_3 orientations are all parallel.

We now have a manifold with a sense of direction. We say that the locally varying A_3 phase has induced an affine connection into the emergent expectation space. What was local variation of A_3 orientation has now become intrinsic curvature within the underlying space.

Above, we saw that the forces like electro-magnetism and the weak nuclear force arise from local variation of the phase of a spinor space. We now have the curvature of our 4-dimensional space-time arising from the same local variation of phase, but, because there was no established affine connection (sense of direction) in the manifold against which the local variation of phase could be measured, that local variation of phase has become embedded in the underlying space as the locally varying curvature of the underlying space.

Of course, within general relativity, curvature of our 4-dimensional space-time is associated with the gravitational force. The gravitational force has emerged as a consequence of locally varying orientation of the A_3 spinor spaces.

The reader will recall that, according to QFT, the other physical forces arose as a consequence of locally varying orientation of the other spinor spaces.

Chapter 13

Some Bits

This book was written for the lay person. As such, we have largely avoided the mathematics of empty spaces. In this penultimate chapter, we get a little technical; we do it in a way that continues to avoid the 'hard mathematics', but we have to use 'hard' terminology. It is hoped that the lay person will read this chapter, but, unless the lay person has a good knowledge of physics, parts of this chapter will seem obscure. The final chapter is much easier than this chapter.

Classical electro-magnetism and GEM:
Although we have not previously mentioned it, classical electro-magnetism emerges from the mathematics of the super-imposition of the A_3 spaces. Classical electro-magnetism emerges as the anti-symmetric part of the emergent expectation A_3 field tensor. The symmetric part of the emergent expectation A_3 field tensor is the gravitational field tensor identified with the energy-momentum tensor.

The Maxwell equations of electro-magnetism emerge from the mathematics of the super-imposition of the A_3 spaces as the anti-symmetric part of the emergent expectation field equations. As the Maxwell equations emerge, they bring with them a classical universe that is entirely without anti-matter. Anti-matter exists only within the spinor spaces.

A similar set of Maxwell-like field equations emerge as the symmetric part of the emergent expectation field equations. These equations would form gravito-electro-magnetism, GEM, except that they exactly cancel each other leaving only curvature as the gravitational field.

Thus, the whole of classical physics emerges from the super-imposition of the A_3 spinor spaces.

Meanwhile, back in quantum physics:

Since the symmetric parts of the A_3 spaces are closely associated with gravity, we believe they are associated with mass. There are three pairs of A_3 spaces, and so we might expect every particle to have three masses within quantum physics. (This will not be the case in classical physics because the A_3 spaces are super-imposed together.) We do observe that every fermion has three masses. This phenomenon is called the three generations of particles.

And super-symmetry:

Within QFT, there are two types of particles. One type of particles are called fermions, and the other type of particles are called bosons.

Fermions are matter particles like the electron. Fermions are spin $\frac{1}{2}$ particles. This means that they rotate in 2-dimensional space-time and in 2-dimensional Euclidean space (that is transform under a Lorentz transformation) in a way that is associated with double cover rotation (like the quaternions or the A_3 spinor spaces).

Bosons are force carrying particles like the photon or the $\{W^\pm, Z^0\}$ particles. Bosons are spin 1 particles which means that they rotate in 2-dimensional space-time and in 2-dimensional Euclidean space (that is transform under a Lorentz transformation) in a way that is associated with single cover (normal) rotation.

The theory of super-symmetry postulates that for every fermion there should be a corresponding 'twin' boson and that for every boson there should be a corresponding 'twin' fermion. The two particles, one a fermion and one a boson, of each set of 'twin' particles should have the same mass.

If the postulate of super-symmetry is correct, then the Higgs boson would have a mass of circa 120 to 130 GeV and would be observable by the Large Hadron Collider at CERN. If the postulate of super-symmetry is not correct, then the Higgs boson would have a mass of millions of TeV and would not be observable by the Large Hadron Collider at CERN. A Higgs boson of mass 125 – 127 GEV has been observed at CERN.

If the postulate of super-symmetry is correct, then the coupling constants (the strengths) of all fundamental forces would be exactly equal at high energies rather than being almost, but not quite, equal. Such exact equality would be unification of the forces of nature and much 'prettier' than a very near miss.

The observation of the Higgs boson at CERN is taken to mean that the postulate of super-symmetry must be correct. However, there is a problem. None of the 'twin' particles, called super-particles or sparticles, has ever been observed. The standard excuse for this is that there must be something not symmetrical about super-symmetry that leads to the super-particles being enormously heavy and beyond reach of our observations even though super-symmetry says that they should be of the same mass as the observed particles. No-one understands what has gone wrong. However, our newly developed understanding of empty space does provide an answer; this might or might not be the correct answer.

QFT considers one of the 2-dimensional space-time rotations in our 4-dimensional space-time to be of the form:

$$\begin{bmatrix} \cosh \chi & \sinh \phi & 0 & 0 \\ \sinh \chi & \cosh \chi & 0 & 0 \\ 0 & 0 & 1 & 0 \\ 0 & 0 & 0 & 1 \end{bmatrix} \quad (13.1)$$

The other two 2-dimensional space-time rotations in our 4-dimensional space-time are similar, but the trigonometric functions are positioned slightly differently within the 4×4 matrix.

QFT considers a 2-dimensional Euclidean rotation in our 4-dimensional space-time to be of the form:

$$\begin{bmatrix} 1 & 0 & 0 & 0 \\ 0 & \cos\theta & \sin\theta & 0 \\ 0 & -\sin\theta & \cos\theta & 0 \\ 0 & 0 & 0 & 1 \end{bmatrix} \quad (13.2)$$

The other two 2-dimensional Euclidean rotations in our 4-dimensional space-time are similar, but the trigonometric functions are positioned slightly differently within the 4×4 matrix.

In both the above cases, (13.1) & (13.2), these are single cover 2-dimensional spinor rotations in our 4-dimensional space-time. Remember, single cover rotation goes only clockwise or only anti-clockwise never both together.

This single cover type of rotation is the only kind of rotation used in QFT. The double cover type of rotations which we have seen in the $C_2 \times C_2$ spinor spaces is not used in QFT; this is perhaps because such double cover rotation matrices are unknown to QFT theorists.

And so, conventional super-symmetry postulates one type of Lorentz transformation (single cover rotation) of the forms (13.1) & (13.2) and two types of particles.

The reader might guess what is coming next. Looking at the A_3 rotation matrix, (9.1), and choosing a co-ordinate system which sets the two variables $\{c,d\}$ to zero, we have another type of rotation. This is a 4-dimensional 2-dimensional rotation (sorry). It is a double cover rotation:

$$A_{3\,Rot} =$$

$$\begin{bmatrix} \cosh(\lambda) & \dfrac{b}{\lambda}\sinh(\lambda) & 0 & 0 \\ \dfrac{b}{\lambda}\sinh(\lambda) & \cosh(\lambda) & 0 & 0 \\ 0 & 0 & \cosh(\lambda) & -\dfrac{b}{\lambda}\sinh(\lambda) \\ 0 & 0 & -\dfrac{b}{\lambda}\sinh(\lambda) & \cosh(\lambda) \end{bmatrix}$$

$$\lambda = \sqrt{b^2}$$

(13.3)

This is a 4-dimensional spinor space-time rotation. It is a rotation which goes both clockwise and anti-clockwise at the same time – that's double cover.

We have two different types of rotation (Lorentz transformation); these are a double cover (spin $\dfrac{1}{2}$) rotation, (13.3), and a single cover (spin 1), (13.1) & (13.2), rotation. We need only one set of particles to get super-symmetry. We already have that one set of particles.

Whether or not this new type of super-symmetry gives the same results as the old type of super-symmetry is not yet known, but it does solve the absence of super-particles problem. Super-symmetry solves problems regarding the energy of the vacuum and what are known as 'quadratic divergences'. Research in this area is ongoing.

Wot! no gravitons:
Within QFT, every force is associated with a boson. A boson is a particle which transmits the force. It is therefore expected that there will be a boson associated with quantum gravity. Such bosons have been postulated and named gravitons.

Gravitons have never been observed. The standard excuse is that they are very difficult to observe.

We have seen above that the electro-magnetic force, the weak nuclear force, and the strong nuclear force are associated with the locally varying orientation of spinor spaces fixed as a fibre bundle on to our underlying 4-dimensional space-time. We have seen how gravity is associated with curvature induced into our 4-dimensional space-time by the locally varying orientation of the A_3 spinor spaces.

Bosons arise because the orientation of the spinor spaces, as measured against the affine connection within our 4-dimensional space-time, varies from point to point in our 4-dimensional space-time. A graviton should arise by the same mechanism. However, since the affine connection of our 4-dimensional space-time is defined by the orientation of the A_3 spaces, as measured against the affine connection within our 4-dimensional space-time, there is no local variation of orientation of the A_3 spinor spaces. And so, there will be no gravitons. This fits with observation.

Breakdown of parity:
When the mathematical objects like field equations or distance functions of spinor spaces are super-imposed, regardless of how many spinor objects are super-imposed, we are left with only one emergent expectation object. When the classical Maxwell field equations of electro-magnetism emerge from the spinor field equations, we are left with one set of field equations. We have not shown it is this book, but, at the time of emergence, there is a choice between whether to format the emergent expectation field equations in a quaternion format, which fits the usual arbitrary definitions of electric field etc., or to format the emergent expectation field equations in an anti-quaternion format, in which case we would have to change our arbitrary definitions. The choice we make is arbitrary, but, having made the choice, we have 'lost' the other half of the spinor spaces. This is how we 'lose' anti-matter in the classical universe.

The quaternions have commutation relations which are clockwise. The anti-quaternions have the same commutation relations but in reverse, that is anti-clockwise. If we have chosen the quaternion format, then the weak force associated with the quaternions will be apparent to us in the classical universe but the weak force associated with the anti-quaternions will not be apparent to us in the classical universe. We think this might be the breakdown of parity. Research is ongoing in this area, but we think this might be why the weak nuclear force is left-handed.

Chapter 14

Concluding Remarks

Perhaps the reader found the previous chapter a little too specialised. We now return to the basics.

There is nothing more than simple mathematics needed to produce all the spinor spaces, that is all the different types of numbers.

We have introduced the super-imposition operation which is no more than a generalisation of the well-established procedure in quantum physics of taking expectation values.

Of all the different types of numbers, only two 2-dimensional types of numbers (2-dimensional spinor spaces) and two 3-dimensional types of numbers (3-dimensional spinor spaces) survive super-imposition with their algebraic structures intact; in particular, with their rotations and angles and distance functions intact.

There is only one emergent expectation distance function, the A_3 emergent expectation distance function, which will support both types of 2-dimensional spinor spaces. This is the distance function of our 4-dimensional space-time. Although we have not shown it in detail in this book, the emergence of the A_3 emergent expectation space brings with it the mathematical apparatus of Riemann geometry and 4-dimensional space-time curvature. In the theory of general relativity, the curvature of 4-dimensional space-time is associated with the gravitational force.

We are unaware of any emergent expectation distance function that will support the 3-dimensional spinor spaces. There might be one, we still have not solved the dark matter problem.

Two types of physics:
Since we have two general types of empty space, spinor spaces and emergent expectation spaces, we might expect to have two types of physics.

One type of physics exists in purely spinor spaces; we think this is quantum physics.

The other type of physics exists in the space that is the unique A_3 emergent expectation space that is our 4-dimensional space-time[28]; the physics of this space we call classical physics.

There is no mathematical connection between the two types of empty space and thus between the two types of physics. The emergent classical universe is formed from spinor spaces that were smashed to bits by the super-imposition operation. The forming of the classical universe is more a gathering together of wreckage than it is a mathematical structure. Technically, we say that the A_3 emergent expectation space is not a division algebra space.

---

[28] The 'unique includes the quaternion emergent expectation space.

Concluding Remarks

Because there is no mathematical connection between the two separate parts of physics, the physics of the spinor spaces and the physics of the classical emergent expectation universe, there is no deterministic connection between the two types of physics. What happens in the spinor spaces seems to be random to we observers sitting in the classical universe. This is why we find probability within quantum physics.

A farewell:
The research into the nature of empty space which we have put before the reader in this book is very recent and still white hot from the furnace. Over the coming years, it will hammered and wrought into a colder and more shapely form. Impurities and misinterpretations will be beaten out of it; gaps of understanding and holes within it will be filled. There is no doubt that the understanding of empty space put before the reader in this book will change over the next few decades. None-the-less, the reader is, at a lay person's level, now very well acquainted with the very forefront of our understanding of empty space.

A philosopher's cogitation:
Philosophers always have the last word.

The philosophically fascinating aspect of all the above is that we have deduced the whole of our classical universe, and hope to eventually deduce the whole of the quantum universe, from nothing more than the real numbers and the finite groups.

From Where Comes the Universe?

In fact, the finite groups are very easily deduced as nothing more than permutations of different whole real numbers.

Further, as shown by Bertrand Russell a century ago, all the real numbers can be deduced from nothing more than the number 1. In short, if the number 1 exists, then the universe exists. Indeed, the whole universe is nothing more than the solitary number 1.

Note: It does not have to be the number one. Any whole number will suffice, including the number 42.

From where comes the universe?
It comes from the number one.

It has been a pleasure writing for you.

Dennis Morris

Port Mulgrave

September 2015

Other Books by the Same Author

The Naked Spinor – a Rewrite of Clifford Algebra

Spinors exist in Clifford algebras. In this book, we explore the nature of spinors. This book is an excellent introduction to Clifford algebra.

Complex Numbers The Higher Dimensional Forms – Spinor Algebra

In this book, we explore the higher dimensional forms of complex numbers. These higher dimensional forms are connected very closely to spinors.

Upon General Relativity

In this book, we see how 4-dimensional space-time, gravity, and electromagnetism emerge from the spinor algebras. This is an excellent and easy-paced introduction to general relativity.

From Where Comes the Universe

This is a guide for the lay-person to the physics of empty space.

Empty Space is Amazing Stuff – The Special Theory of Relativity

This book deduces the theory of special relativity from the finite groups. It gives a unique insight into the nature of the 2-dimensional space-time of special relativity.

The Nuts and Bolts of Quantum Mechanics

This is a gentle introduction to quantum mechanics for undergraduates.

Quaternions

This book pulls together the often separate properties of the quaternions. Non-commutative differentiation is covered as is non-commutative rotation and non-commutative inner products along with the quaternion trigonometric functions.

The Uniqueness of our Space-time

This book reports the finding that the only two geometric spaces within the finite groups are the two spaces that together form our universe. This is a startling finding. The nature of geometric space is explained alongside the nature of division algebra space, spinor space. This book is a catalogue of the higher dimensional complex numbers up to dimension fifteen.

Other Books by the Same Author

Lie Groups and Lie Algebras

This book presents Lie theory from a diametrically different perspective to the usual presentation. This makes the subject much more intuitively obvious and easier to learn. Included is perhaps the clearest and simplest presentation of the true nature of the Lie group $SU(2)$ ever presented.

The Physics of Empty Space

This book presents a comprehensive understanding of empty space. The presence of 2-dimensional rotations in our 4-dimensional space-time is explained. Also included is a very gentle introduction to non-commutative differentiation. Classical electromagetism is deduced from the quaternions.

The Electron

This book presents the quantum field theory view of the electron and the neutrino. This view is radically different from the classical view of the electron presented in most schools and colleges. This book gives a very clear exposition of the Dirac equation including the quaternion rewrite of the Dirac equation. This is an excellent introduction to particle physics for students prior to university, during university and after university courses in physics.

The Quaternion Dirac Equation

This small book (only 40 pages) presents the quaternion form of the Dirac equation. The neutrino mass problem is solved and we gain an explanation of why neutrinos are left-chiral. Much of the material in this book is drawn from 'The Electron'; this material is presented concisely and inexpensively for students already familiar with QFT.

An Essay on the Nature of Space-time

This small and inexpensive volume presents a view of the nature of empty space without the detailed mathematics. The expanding universe and dark energy is discussed.

Elementary Calculus from an Advanced Standpoint

This book rewrites the calculus of the complex numbers in a way that covers all division algebras and makes all continuous complex functions differentiable and integrable. Non-commutative differentiation is covered. Gauge covariant differentiation is covered as is the covariant derivative of general relativity.

Even Mathematicians and Physicists make Mistakes

This book points out what seems to be several important errors of modern physics and modern mathematics. Errors like the misunderstanding of rotation, the failure to teach the higher dimensional complex numbers in most universities, and the mathematical inconsistency of the Dirac equation and some casual errors are discussed. These errors are set in their historical circumstances and there is discussion about why they happened and the consequences of their happening. There is also an interesting chapter on the nature of mathematical proof within our society, and several famous proofs are discussed (without the details).

Finite Groups – A Simple Introduction

This book introduces the reader to finite group theory. Many introductory books on finite groups bury the reader in geometrical examples or in other types of groups and lose the central nature of a finite group. This book sticks firmly with the permutation nature of finite groups and elucidates that nature by the extensive use of permutation matrices. Permutation matrices simplify the subject considerably. This book is probably unique in its use of permutation matrices and therefore unique in its simplicity.

The Left-handed Spinor

This book covers the left-handed parts of mathematics which we call the chiral algebras. These algebras have CP invariance, violation of parity, and many other aspects which makes them relevant to theoretical physics. It is quite a revelation to discover that mathematics is left-handed.

Non-commutative Differentiation and the Commutator

(The Search for the Fermion Content of the Universe)

From Where Comes the Universe?

This book develops the theory of non-commutative differentiation from the fundamentals of algebra. We see what an algebraic operation (addition, multiplication) really is, and we discover that the commutator is a third fundamental algebraic operation within some division algebras. This leads to the first part of the derivation of the fermion content of the universe.

Index

1

1-dimensional numbers, 16
1-dimensional space, 9, 13

2

2-dimensional curl, 74
2-dimensional distance function, 97
2-dimensional inner product, 74
2-dimensional numbers, 16, 20, 21, 58
2-dimensional rotation, 30, 37
2-dimensional space, 9, 20
2-dimensional space-time numbers, 38
2-dimensional space-time spinor rotation, 92

3

3-dimensional angle, 96
3-dimensional numbers, 41, 55
3-dimensional rotation, 55
3-dimensional rotation matrix, 55
3-dimensional space, 86
3-dimensional trigonometric functions, 55
3-dimesional spinors space, 96

4

4-dimensional angle, 44
4-dimensional inner product, 81
4-dimensional numbers, 40, 58
4-dimensional rotation, 43, 76
4-dimensional rotation matrix, 75
4-dimensional space-time, 33
4-dimensional spinor space, 96
4-dimensional spinor space-time rotation, 114

8

8-dimensional Lie group, 93
8-dimensional spaces, 85
8-dimensional spinor spaces, 83, 94

A

accelerated reference frame, 98, 99
action at a distance, 88
affine connection, 81, 95

affine connection, induced, 107
algebraic structure destroyed, 76
angle, multi-argument, 36
anti-matter, 110
Argand, 24
Argand diagram, 24
Ars Magna, 22
axioms of a division algebra, 18
axis of rotation, 35

B

bosons, 110
breakdown of parity, 116

C

Cardano, 22, 23
Cartesian co-ordinates, 24
CERN, 111
classical electromagnetism, 109
classical physics, 4, 9
Clifford algebras, 18, 46, 76
Cockle, 38
collapse of the wave-function, 82
colour force, 95
commutation relations, 92, 93
commutativity, 18
complex number in polar form, 28
complex numbers, 21
coshine function, 38, 65
cosine function, 26
cosine function, graph, 27
coupling constants, 111

D

dark matter, 9
defining properties, 18
defining properties of numbers, 18, 42
determinant, 61
determinant of a matrix, 21
determinism, 5, 72
deterministic connection, 119
distance function, 22, 79
distance function, 2-dimensional space-time, 40
distance function, space-time, 33
distant galaxies, 70
division algebra, 18, 21
double cover, 45
double cover rotation, 111, 114
double rotation, 47

E

Einstein, 37, 98
electromagnetic field, 91
electron, 110
electron spin, 35
electron, rotation through $720°$, 47
emergent expectation field equations, 116
emergent expectation field tensor, 109
emergent expectation space, 84, 85
emergent manifold, 77
empty space, 5, 8

energy-momentum tensor, 109
equivalence principle, 98, 99
Euclidean complex numbers, 21
Euclidian complex number, polar form, 28
Euler, 35
expanding universe, 70
expectation algebra, 76
expectation distance function, 61, 78, 85
expectation value, 72
exponential function, 39

F

fabrication, 81, 87
fermions, 110
Ferrari, 22
fibre bundle, 32, 88, 89
field, 88
field space, 89
finite groups, 9, 52, 56
flat manifold, 77
flat tangent space, 79
folding, 94
force as a change of phase, 92
fractions, 12
free fall, 98
fundamental forces, 111

G

Gauss, 24
GEM, 110
general relativity, 78, 87, 98, 108
general theory of relativity, 63
globally flat manifold, 78
gluon, 95
GR, 98
GR, assumptions, 101
GR, field equations, 99, 102
GR, postulates, 98
gradient, 37
grand unification, 111
gravitation as a scalar field, 103
gravitation as a tensor field, 103
gravitation as a vector field, 103
gravitation, Newton, 98
gravitational force, 108
gravito-electromagnetism, 110
gravitons, 115

H

Hamilton, 42
Higgs boson, 111
hyperbola, 68
hyperbolic circle, 68
hyperbolic complex numbers, 38, 63, 91
hyperbolic trigonometry, 65

I

imaginary numbers, 13, 23
imaginary variables, 76
inertial mass, 92
inertial reference frame, 98
intrinsic curvature, 99
intrinsic spin, 36

intrinsic spin of an electron, 30
irrational numbers, 14
isomorphic division algebras, 73
isotropy of 2-dim space-time, 63
isotropy of space, 62

K

Kepler, Johannes, 4

L

Large Hadron Collider, 111
length of the number, 22
Lie groups, 89, 93
local flatness, 78
local variation of phase, 91, 108
local variation of quaternion phase, 93
locally flat manifold, 78
Lorentz transformation, 56, 111, 112
Lorentz transformation, spinor, 113

M

magnetic field, 64
manifold, 77, 106
matrix, 20
Maxwell equations, 109
metric tensor, 81
multiplication operation destroyed, 76
multiplicative closure, 17, 21
multiplicative commutativity, 41
multiplicative inverse, 18
multiplicative non-commutativity, 41

N

negative numbers, 12
Newton, Isaac, 4, 32, 98
non-commutative division algebra, 42
non-locality, 82
nu-functions, 55
numbers, 11

O

only possible classical universe, 86
orientation, 96, 107

P

parallel lines, 77, 106
parity, 116
permutation matrices, 53
permutations, 52
phase of the wave function, 91
photon field, 90
Plancks constant, 102
polar co-ordinates, 25
possible 4-dimensional distance functions, 83
Pythagoras theorem, 22, 33, 40
Pythagoras theorem, space-time, 40, 65

Index

Q

QFT, 4, 88
quantum field, 89
quantum field theory, 4, 88
quantum mechanics, 82
quantum physics, 4, 72
quaternion matrix, 41
quaternion space, 93
quaternion trigonometric functions, 43
quaternions, 40, 82

R

rational numbers, 12
real number line, 12
real numbers, 12
Riemann distance function, 79
Riemann geometry, 78, 105
Riemann space, 103, 105
rotation, 35
rotation matrix, 29, 30, 31, 37
rotation matrix of space-time, 38
rotation matrix, quaternion, 43
rotation not about an axis, 30
rotation, 2-dim space-time, 37
rotation, 2-dimensional, 24
rotation, double cover, 45

S

sense of direction, 77, 95, 106
sense of direction, induced, 107
shine function, 38, 65
simple finite groups, 56
simplicity, 6
sine function, 26
space and time, Newtonian, 32
space-time trigonometric functions, 38
special relativity, 87
special theory of relativity, 40, 56, 62
spin half particles, 110
spin one particles, 111
spinor rotation, 35, 43, 49, 57
spinor spaces, 117
sporadic groups, 57
straight line, 2-dimensional space-time, 69
strong nuclear force, 93, 95
super-imposition, 73
super-imposition, 2-dimensional spinors, 74
super-particles, 112, 114
super-symmetry, 111
surviving super-imposition, 74

T

tangent space, 78
Tartaglia, 22
three generations of particles, 110
time, 36
time dilation, 67, 69
trigonometric functions, 38

trigonometric functions, infinite sums, 39
trigonometry, 25
two types of physics, 118

U

unification, 6
unification, of physics, 5

V

velocity, 37, 40

W

Wallis, 23
waves, 28
weak nuclear force, 82, 93
weak nuclear force is left-handed, 116
Wessel, 24
whole numbers, 11

Z

zero divisors, 18

Made in United States
Orlando, FL
20 March 2022